KT-210-360

Forestry Commission

Field Book 8

The Use of Herbicides in the Forest 1995

by Ian Willoughby

Silviculturist
The Forestry Authority
Research Division
Alice Holt Lodge
Wrecclesham Farnham
Surrey GU10 4LH

and Jim Dewar

Technical Development Forester
The Forestry Authority
Technical Development Branch
Ae Village, Dumfries
DG1 1QB

LONDON: HMSO

© Crown copyright 1995
Fourth edition 1995
First edition Crown copyright 1983
Applications for reproduction should be made to HMSO Copyright Unit

ISBN 0 11 710330 6
ODC 414:441:236.1:307

Keywords: *Herbicides, Forestry*

Enquiries relating to this publication should be addressed to:
The Research Publications Officer
The Forestry Authority, Research Division
Alice Holt Lodge, Wrecclesham
Farnham, Surrey GU10 4LH

Acknowledgements

Thanks are due to all those who made suggestions, provided information, and commented on the text, in particular the herbicide manufacturers, Forestry Authority Education, Safety & Training staff, and to John Morgan, David Clay, Richard Toleman, David Williamson and John Budd. Andy Moffat provided Section 3.6.2. Graphics were prepared by John Williams. Translation of handwriting into typed draft was by Pam Wright.

This Field Book is based substantially on an earlier edition by Williamson and Lane (1989).

Disclaimer

This Field Book is intended as a guide to the use of herbicides in the forest – it is published to advise Forestry Commission staff but, as a service, it is made available to the forest industry. It is not intended as an endorsement or approval by the Forestry Commission of any product or service to the exclusion of others that may be available. The book is, to the author's knowledge, as accurate and up-to-date as possible. However, the Forestry Commission can accept no responsibility for any loss or damage resulting from herbicide applications, or from following any advice in this book.

Preface

This edition of *The use of herbicides in the forest* has been prepared as a successor to the 1989 title of the same name, by Williamson and Lane. The Field Book retains the same layout and general contents of previous editions, but is updated to take into account herbicide and applicator innovations, as well as commercial and statutory withdrawal of products. The major difference from previous editions is that all herbicides with full forestry approval are included. Products which are imminently to be withdrawn at the time of writing are not treated in detail but are summarised at the end of the book. A comprehensive list of all products with forestry approval, whether referred to in detail or not, is also given. In addition, there is a section on the use of herbicides in farm forestry situations.

If used correctly, herbicides can be of considerable assistance in establishing new woodlands and in re-stocking existing ones. This Field Book aims to assist managers in safe and effective vegetation management.

Contents

1 Introduction

1.1General information 1
1.2 Content and layout 3
1.3 Nomenclature of herbicides 3
1.4 Assumptions and conventions used in this Field Book 4
1.4.1 Area 4
1.4.2 Crop tolerance 4
1.5 Application patterns 5
1.6 Application methods 5
1.7 The decision chain – a summary 7
1.8 Costs 8
1.9 References for further reading 9

2 Pesticide legislation

2.1 Introduction 13
2.2 Products 13
2.3 Conditions of approval 14
2.4 Minor uses 14
2.5 Operators 17
2.6 Use of adjuvants 19
2.7 Aerial application of pesticides 19
2.8 Codes of practice 19
2.9 Poisons Act 1972 20
2.10 The Poisonous Substances in Agriculture Regulations 1984 20

3 Safety precautions and good working practices

3.1 Private sector training 21
3.2 The Forestry and Arboriculture Safety and Training Council Safety Guides 21
3.3 Routine precautions 22
3.4 Storage and transport of herbicides 22

3.5 Herbicides and the environment 23
3.6 Use of herbicides in or near water 24
3.6.1 Avoiding water contamination 24
3.6.2 Leaching and run-off risks associated with herbicide applications 27
3.7 Tank mixes 31
3.8 Reduced volume applications 31
3.9 Safe disposal of surplus herbicide 32
3.9.1 Unused dilute herbicide 32
3.9.2 Surplus herbicide concentrate 33
3.9.3 Old or deteriorated herbicide 33
3.9.4 Contaminated solid material 34

4 Grasses, herbaceous broadleaved weeds and mixtures

4.1 General 35
4.2 Atrazine 59
4.3 Clopyralid 70
4.4 Dalapon/Dichlobenil 75
4.5 2,4-D 77
4.6 2,4-D/Dicamba/Triclopyr 79
4.7 Glufosinate ammonium 81
4.8 Glyphosate 83
4.9 Imazapyr 87
4.10 Isoxaben 89
4.11 Propyzamide 91
4.12 Triclopyr 94

5 Bracken

5.1 General 97
5.2 Asulam 98
5.3 Dicamba 101
5.4 Glyphosate 103
5.5 Imazapyr 107

6 Heather

6.1 General 109
6.2 2,4-D 111
6.3 Glyphosate 114
6.4 Imazapyr 118

7 Woody weeds

7.1 General 120
7.2 Types of treatment 121
7.3 Foliar treatment 124
7.3.1 2,4-D 124
7.3.2 2,4-D/Dicamba/Triclopyr 126
7.3.3 Fosamine ammonium 128
7.3.4 Glyphosate 130
7.3.5 Imazapyr 134
7.3.6 Triclopyr 136
7.4 Stem treatment 139
7.4.1 Ammonium sulphamate 139
7.4.2 Glyphosate 141
7.4.3 Triclopyr 144
7.5 Cut stump treatment 146
7.5.1 Ammonium sulphamate 146
7.5.2 2,4-D/Dicamba/Triclopyr 148
7.5.3 Glyphosate 150
7.5.4 Triclopyr 153

8 Rhododendron

8.1 General 155
8.2 Ammonium sulphamate 156
8.3 2,4-D/Dicamba/Triclopyr 158
8.4 Glyphosate 160
8.5 Imazapyr 163
8.6 Triclopyr 166

9 Farm forestry weed control only

9.1 General 168
9.2 Cyanazine 178
9.3 Fluazifop-p-butyl 186
9.4 Metazachlor 197
9.5 Pendimethalin 202
9.6 Propaquizafop 206

10 Protective clothing and personal equipment

10.1 General 208
10.2 List of recommended products and suppliers 209
10.3 Cleaning recommendations 210
10.4 Protective clothing and equipment recommendations 212

11 Equipment

Detailed list of contents of Section 11 215
11.1 Disclaimer 215
11.2 Volume rate categories 216
11.3 Calibration 217
11.4 Nozzles 221
11.5 Applicators for liquid herbicides 222
11.6 Applicators for spot application of granules 258
11.7 Tree injection applicator 260
11.8 ENSO brush-cutter stump treatment attachment 261
11.9 Pesticide transit box 262
11.10 Output guides – herbicide application 262
11.11 Output guide – overall or band application 263
11.12 Output guide – spot application 271

12 Lists of herbicides and manufacturers

12.1 Products with full forestry label approval 277
12.2 Products with full farm forestry label approval 278
12.3 Products with forestry off-label approval 278
12.4 Products with farm forestry off-label approval 278

12.5 Herbicides used in forestry which have approvals expired or have been commercially withdrawn during the last two years 279
12.6 Products with approval for use in or near water for aquatic weed control 280
12.7 Additives mentioned in the text 281
12.8 Product suppliers 281

13 Sources of advice

13.1 Forestry Authority offices 288
13.2 Other sources of advice 293

14 Glossary and abbreviations

14.1 Abbreviations used in text 294
14.1.1 Species 294
14.1.2 Other abbreviations and symbols 294
14.2 Glossary of general and technical terms 295

15 Index to weeds and chemicals 305

List of tables

Table number	Title	Page
1	Run-off and leaching ratings for approved forestry herbicides	28
2	Relative soil vulnerability to run-off and leaching	29
3	Determining risk from run-off and leaching	30
4	Susceptibility of common grasses in the forest to approved herbicides	38
5	Susceptibility of herbaceous vegetation to approved herbicides	43
6	Product rates for woody weeds susceptible to foliar applications	122
7	Crop tolerance to farm forestry herbicides	170
8	Susceptibility of common arable weeds to selective farm forestry herbicides	172
9	Farm forestry herbicide tank mixes	177
10	Protective clothing and equipment recommendations	212
11	The proportion of 1 hectare of plantation treated during spot applications	220
12	The proportion of 1 hectare of plantation treated during band applications	220
13	Nozzle output data – CP15 knapsack sprayer	223
14	Nozzle output data, when using a knapsack sprayer fitted with a spray management valve	227
15	Volume rates for the dribble bar applicator	236
16	Nozzle output for the Fox Motori Electra F11OTS knapsack sprayer	239
17	Quantity of herbicide product required per 5 litre knapsack – Forestry spot gun	244
18	Required nozzle output for 1.2 metre swathe – Microfit Herbi	246
19	Nozzle output for ULVA 8	250
20	Required nozzle output for 2 metre swathe – ULVA 8	252

1 Introduction

1.1 General information

The two most important pieces of legislation concerning the use of herbicides are the Control of Pesticides Regulations 1986 and the Control of Substances Hazardous to Health Regulations 1988. All recommendations in this Field Book should comply with this legislation.

This Field Book contains advice on the use of herbicides in the forest. In this context, 'forestry' includes vegetation management amongst newly established trees, in restock areas of existing woodland, in shelterbelts, and other areas within forests such as on ride-side edges, for conservation purposes. All the products in this Field Book have either full forestry or farm forestry approval, off-label approval for use in forestry, or off-label approval for use in farm forestry. At the time of writing only the products listed have approval. It is not permissible to use a product with the same active ingredient as an approved product, unless a forestry approval exists on the label. Check the product label if there is any doubt.

Products with label or off-label approval for use in farm forestry, i.e. land previously under arable cultivation or improved grassland that is being converted to farm woodland or short rotation coppice, cannot be used in other forestry situations.

Not all of the products listed have been subject to Forestry Commission testing or experimentation. Entries in these cases are based on manufacturer's label information – product information has been verified by experimental results submitted to the Pesticides Safety Directorate, MAFF.

Products with approval for amenity or woody ornamental situations have not been included – certain users may be able legally to use these products depending on their situation. Further information can be gained from *The UK pesticides guide 1995* (CAB International/BCPC) and individual product labels.

IT IS ESSENTIAL THAT USERS REFER TO INDIVIDUAL PRODUCT LABELS BEFORE ANY APPLICATIONS OF HERBICIDE ARE MADE.

The Field Book has been written to comply with the codes of practice on the use of herbicides in the forest. HOWEVER, ALL USERS MUST REFER TO THE HSE PUBLICATION *THE SAFE USE OF PESTICIDES FOR NON-AGRICULTURAL PURPOSES* BEFORE EMBARKING ON ANY PESTICIDE APPLICATIONS. Users should also consult Forestry Commission Occasional Paper 21 *Provisional code of practice for the use of pesticides in forestry* as, although this publication has been superseded by the HSE guidance, it contains useful additional information on environmental protection, and on the safe storage and transport of pesticides (see Section 1.9).

The application equipment described in this Field Book is a representative sample of the various applicators which the Forestry Commission's Technical Development Branch have found to be effective.

This Field Book is primarily concerned with the use of herbicides or weedkillers. Other chemicals such as insecticides, fungicides, mammalian poisons and repellents are, together with herbicides, collectively referred to as pesticides. Pesticides other than herbicides are only referred to in Sections 2 and 3 dealing with the relevant legislation which affects the use of all the chemical products in this field. For detailed guidance on using pesticides other than herbicides, users should consult with the Entomology, Pathology or Woodland Ecology Branches at the Forestry Commission's Northern or Southern research stations. Research Information Note 185 *Approved methods for insecticidal protection of young trees against Hylobius abietis and Hylastes species* may also be of use.

Recommendations for weed control in forest nursery seed beds and transplant lines can be found in Forestry Commission Technical Paper 3 *Forest nursery herbicides* and in Forestry Commission Bulletin 111 *Forest nursery practice.*

This Field Book will be revised as frequently as is necessary to keep it up-to-date and useful as a working manual. Between editions, updating information will be made available through Research Information Notes, published by the Forestry Commission, Research Division, Alice Holt Lodge, Wrecclesham, Farnham, Surrey GU10 4LH.

1.2 Content and layout

The main sections of this Field Book (Sections 4–8) are laid out by reference to major weed types: grass/herbaceous broadleaved, bracken, heather, woody weeds, and rhododendron. The herbicides appropriate for use against this range of forest weed species are set out in the accompanying wallchart and Figures 1–3. These can be used to help identify potential herbicides. Once candidate herbicides have been selected, detailed information on particular products can be obtained from the relevant section of the Field Book.

Before making a final selection of any herbicide, managers must consider whether there are other more appropriate means of achieving their objectives than by use of a herbicide.

Herbicide entries are set out in a standard format:

- description and properties of forestry approved products, crop tolerance
- recommended application rates
- methods and timing of application
- additional information on weed control
- protective clothing and special precautions

Sections 2 and 3 on legislation and safe working practices, and Sections 10 and 11 on protective clothing and application equipment, contain very important information upon which the safety and effectiveness of any herbicide treatment depends. PLEASE READ THEM CAREFULLY AND FOLLOW THEIR GUIDANCE whenever you undertake a programme of weed control with herbicides.

1.3 Nomenclature of herbicides

Herbicides can be referred to by one of the following different types of name: e.g. for *glyphosate.*

Product name:

Roundup – the name registered by the manufacture for a product containing a specific formulation of an active ingredient (a.i.). Approval is sought and given for the product by name.

Common name:

glyphosate – the accepted short name for an active ingredient.

Chemical name:

N-phosophonomethyl glycine; the full scientific name for the active ingredient.

These terms could be combined as follows: Roundup is a liquid formulation of glyphosate, containing 360 grams per litre (g/l) of an active ingredient (N-phosphonomethyl glycine) as the isopropylamine salt.

In this Field Book only the product name or common name is used.

1.4 Assumptions and conventions used in this Field Book

(see also Section 14 for glossary and list of abbreviations)

1.4.1 Area

Throughout this Field Book, unless the context clearly indicates otherwise, all references to areas (usually hectares) refer to TREATED AREA, that is the area of ground or plantation that is actually covered with herbicide, i.e. the total area of spots or bands treated within a plantation.

1.4.2 Crop tolerance

Crop tolerance is described using the following terms:

tolerant

unaffected or undamaged by exposure to a herbicide;

moderately tolerant

some damage or check from a herbicide is likely, but the crop species should survive;

sensitive

easily damaged by a herbicide – death of crop species is possible.

The descriptions of crop tolerance assume average site conditions and healthy crop trees (prior to treatment). Not only is the crop at risk if an overdose is applied but under the Control of Pesticides Regulations, it is illegal to apply a rate of pesticide greater than that stated on the label for the use of that pesticide.

For each herbicide entry, where appropriate, there is a note of any necessary waiting period after treatment with the herbicide and before it is safe to plant on the site.

The manager should bear in mind that crop tolerance and weed susceptibility can be affected by site, crop dormancy, crop health, provenance, weather, season, etc. They should proceed with caution in choosing herbicides and rates until confident of the efficacy of treatment in the local conditions within which they are working.

1.5 Application patterns

A herbicide can be applied in a number of ways:

- overall – the herbicide is applied over the whole weeding site;
- band – the herbicide is applied in a band over or between crop trees;
- spot – the herbicide is applied as individual spots around or over each tree.

A directed application is a spray which is directed to hit a target weed and to avoid crop trees.

It is usually easier and quicker, where terrain and crop allow, to apply herbicide as an overall treatment, but this economy of effort imposes a greater risk to the local woodland habitat and uses more herbicide than is strictly required for effective weed control.

When planning pre-planting weed control it is important to consider whether planters will be able to identify treated areas if herbicides are applied in spots or bands. These two techniques may be inappropriate if rhizomatous or stoloniferous weeds such as couch grass are present.

In a post-planting situation where overall application is undesirable, spot or band treatment may be appropriate because usually young trees can be clearly seen.

Directed spot applications are more demanding of time and skill but minimise the amount of herbicide used, the risk to which the crop is exposed and the impact on the environment. They are more easily carried out where trees are protected in treeshelters.

1.6 Application methods

The equipment used to apply a herbicide will depend upon the nature of the herbicide (granular or liquid) and on the application pattern required.

Factors to consider for:

Granular application

a. no water is required
b. no mixing required
c. low cost applicators
d. no problems associated with the disposal of unused diluted pesticide
e. granular applicators require closer monitoring to achieve accurate calibration
f. granular products are often more expensive
g. granular products are bulkier to store and transport
h. there is a more limited range of products

Liquid application

a. water is required except with a few ULV (ultra low volume) products
b. mixing is required except with a few ULV products
c. applicators are more expensive than the simple granular applicators
d. applicator failure or climatic change can lead to the need to dispose of unused diluted pesticide
e. liquid applicators are easier to calibrate accurately, but may require more intensive supervision and training (calibration of the Forestry spot gun is particularly simple)
f. there is a greater choice of liquid products
g. with many products the volume rate of application can be altered to suit various types of applicator
h. liquid or wettable powder formulations are less bulky to store and transport

Section 11 gives detailed guidance on the various types of applicator available.

Aerial application of herbicides is covered by the Control of Pesticides Regulations 1986. Managers wishing to consider this technique must first study the legal obligations under these regulations and should then contact the manufacturer of the product if they wish to consider aerial application further. Only one herbicide (asulam) is currently fully approved for aerial application in forestry situations.

1.7 The decision chain – a summary

The following sequence briefly describes the assessments and decisions involved in achieving correct application of a liquid herbicide. The sequence for a granular herbicide is similar but with fewer variables to reconcile.

Using Sections 4 to 9:

a. From crop and weed characteristics, determine the choice of suitable herbicides, dose rates and application patterns.

b. Make an assessment of the effects of the proposals on operators and the environment ensuring that all necessary equipment is available, safeguards known, and competent and certificated operators available.

c. Consider specifically whether any factors of the locality, operation, etc., add to the risk normally associated with the proposed use.

d. Consider any other factors limiting the choice of herbicide, applicator, droplet size, dilution rate or diluent.

e. Select a suitable:
dose rate
applicator
droplet size (if critical)
application pattern, method and timing*
application rate
CONSULT THE HERBICIDE LABEL

f. Make a COSHH assessment.

Using Section 10:

g. Having decided on the herbicide and method of application, ensure suitable protective equipment is available to operators.

Using Section 11:

h. i. Calculate the likely equipment requirements and settings, choosing values for the relevant variables, e.g. for the knapsack sprayer:
walking speed
nozzle size
swathe width
application rate
dilution

ii. Calibrate the equipment to achieve the correct application rate.

**Note* – the critical period for weed control is from April to June, as this is the time that soil moisture defects are most likely. Whilst herbicide applications may be made at any time of year, this is probably the period during which it is most important to maintain weed-free conditions. The precise period of time that weed-free conditions will need to be maintained on a site will depend on factors such as crop species and spacing, site type, weed species, etc. Essentially, the benefits of weeding must be weighed against the cost of the operation. Generally speaking, weeding should continue until crop trees are the dominant vegetation on the site, and until the likely level of crop-tree suppression or death from weed competition becomes acceptable. A 1 m^2 spot or 1 m wide weed-free band around the tree for 3–5 years after planting is often regarded as an appropriate, economically viable measure. It is usually preferable to attempt to predict the potential weed problems on a site, and choose a herbicide and application method accordingly, rather than waiting and reacting if problems develop. Weeds may be much easier and cheaper to control if this pre-emptive method of weed control is followed, and competition with trees is likely to be less severe than if weeds are left to develop before any action is taken. An example of pre-emptive weed control would be to use a herbicide such as imazapyr pre-planting, to give a degree of residual weed control on a fertile site where a manager anticipates a wide spectrum of prolific weed growth developing. This may well be easier and more cost effective than using repeated, complicated, directed applications of foliar acting herbicides later in the growing season, when weeds have become a serious problem. This approach to weed control is by no means limited to the use of residual herbicides. The use of foliar acting herbicides during the dormant season to control existing weeds, which are likely to grow and compete vigorously with trees in the next growing season, may again be preferable to the use of guarded or directed treatments later in the year. A pre-emptive approach becomes particularly important on more fertile restock sites, and in the conversion of land from agriculture to forestry. In these situations both pre-emptive treatments, and applications during the trees growing season, may be necessary.

1.8 Costs

The cost of any herbicide treatment is a vital consideration for any forest manager. As has been stated earlier, it is generally agreed that a 1 m^2 weed free spot or band around trees for the first 3–5 years after planting is an economically justifiable measure. Herbicide application is often the cheapest way of achieving this.

There are two elements to the cost of any herbicide treatment – the cost of the herbicide, and the cost of applications.

To calculate the cost per hectare of the herbicide, simply multiply the total estimated litres of herbicide required per hectare by their cost per litre.

For overall applications the total herbicide volume required is equal to its product rate per hectare.

For band applications, use Table 12 in Section 11.

For spot applications, use Table 11 in Section 11 to find the proportion of 1 ha treated during spot applications, and multiply this by the product rate to give the amount of product used per hectare.

Sections 11.10-11.12 include two Output Guides that can be used to estimate time taken, and hence cost, of actually applying the herbicide using different application techniques.

Details of additional Output Guides with relevance to weed control are contained in Section 1.9.

1.9 References for further reading

Several references to other published books and leaflets are to be found in the appropriate sections of this Field Book. The following titles will also be useful for background reading and provide further details of some aspects of herbicide practice:

Forestry Commission publications

(available from the Publications Section, Forestry Commission, Research Division, Alice Holt Lodge, Wrecclesham, Farnham, Surrey GU10 4LH. Tel: 01420 22255).

- Aldous, J.R. and Mason, W.L. (eds), Bulletin 111 *Forest nursery practice,* 1994.
- Anon., Occasional Paper 21 *Provisional code of practice for the use of pesticides in forestry,* 1989 (superseded by the HSE *Code of practice for the safe use of pesticides for non-agricultural purposes*).
- Anon., *Nature conservation guidelines,* 1990.
- Anon., *Forests and water guidelines,* 1993.

- Anon., Forestry Practice Guides 1–8 *The management of semi-natural woodlands,* 1994.
- Davies, R.J., Handbook 2 *Trees and weeds – weed control for successful tree establishment,* 1987.
- Edwards, C., Morgan, J. and Willoughby, I., Research Information Note 246 *Approved herbicides for use in forestry,* 1994.
- Edwards, C., Tracy, D.R. and Morgan, J.L., Research Information Note 233 *Rhododendron control by imazapyr,* 1993.
- Nelson, D.G., Research Information Note 187 *Control of Sitka spruce natural regeneration,* 1990.
- Stables, S. and Nelson, D.G., Research Information Note 186 *Rhododendron ponticum control,* 1990.
- Tabbush, P.M., Bulletin 73 *Rhododendron ponticum as a forest weed,* 1987.
- Tracey, D.R. and Nelson, D.G., Research Information Note 203 *Methods of grass control in the uplands,* 1991.
- Williamson, D.R., Research Information Note 201 *Herbicides for farm woodlands and short rotation coppice,* 1991.
- Williamson, D.R., Research Information Note 170 *A brief guide to some aspects of the control of pesticide regulations 1986,* 1990.
- Williamson, D.R., Mason, W.L., Morgan, J.L., Clay, D.V., Technical Paper 3 *Forest nursery herbicides,* 1994.

Forestry Commission Work Study and Technical Development Branch Reports on Pesticide Applicators and Protective Clothing may be obtained from:

Forestry Commission
Technical Development Branch
Ae Village
Dumfries
DG1 1QB
Tel: 01387 860 264

Recent reports include:

Report	*Title*
03/90	*Teejet calibration calculator*
22/90	*Filtering facepiece respirators*
34/90	*Treemate Kerb granule applicator*
36/90	*Avoiding herbicide application errors*

37/90 *Dribble bar*
22/91 *Fox Motori F11OTS knapsack*
7/93 *Oil and chemical spillages*

Information Notes

4/93 *Respiratory protection for herbicide spraying*
7/92 *Rest allowances and protective clothing*
8/93 *The Micron Herbi-4*
8/94 *An introduction to the use of tractor-mounted sprayers in farm woodland.*

The following Output Guides are available at a cost of £2.00 each (July 1993) from:
Forestry Commission
Technical Development Branch
Ae Village, Dumfries, DG1 1QB
Tel: 01387 860 264

- Weeding 1 *Herbicide application*
  Knapsack for overall or band application
  ULVA for overall application
  Microfit Herbi/Herbi for overall or band application
- Weeding 2 *Spot application by spot gun, weedwiper or pepperpot*
- Weeding 3 *Hand tools or portable brush cutter*
- Weeding 5 *Tractor mounted CDA using flowable atrazine*

(*Note:* Elements of Weeding 1 and 2 are reproduced in Section 11.11)

Arboricultural Advisory & Information Service Publications

(available from the AAIS, Alice Holt Lodge, Wrecclesham, Farnham, Surrey GU10 4LH Tel: 01420 22022)

- Hawke, C. and Williamson, D.R. Arboriculture Research Note 106/92/EXT *Japanese knotweed in amenity areas,* 1992.
- McCavish, W.J., Insley, H. (rev. Williamson, D.R.). Arboriculture Research Note 27/92/SILS *Herbicides for sward control among broadleaved amenity trees,* 1992.

Other publications

- *The UK pesticide guide* 1995. CAB International/British Crop Protection Council.
- *Weed control handbook:*
  Principles (Eighth edition, 1989) edited by R.A. Hance and K. Holby.
  Recommendations (Eighth edition, 1978) edited by J.D. Fryer and R.J. Makepeace.

- *The pesticides register* (published monthly). Published by HMSO (Tel: 0171 873 9090).
- *Pesticides 1995*. MAFF Reference Book 500. Pesticides approved under the Control of Pesticides Regulations, 1986.
- *Guidelines for the use of herbicides on weeds in or near watercourses and lakes*. MAFF Booklet 2078 (1985.)
- *Pesticides; guide to the new controls*. MAFF leaflet UL79, 1987. (Available from MAFF Publications, London SE99 7TP.)
- *Working with pesticides guide: the requirements and your responsibilities*. Agricultural Training Board. (Available from the National Agricultural Centre, Kenilworth, Warwickshire. Tel: 01203 696996.)
- *Code of practice of suppliers of pesticides to agriculture, horticulture and forestry*. (Available from MAFF Publications, London SE99 7TP.)
- *Code of practice for the safe use of pesticides on farms and holdings*. (Available from HSE Publications, Tel. 01787 881169.)
- ***Code of practice for the safe use of pesticides for non-agricultural purposes***. (Available from HSE Publications.) **Essential reading**
- *Training recommendations for non-agricultural users of pesticides*. (Available from HSE publications.)
- *Weed guide*. Schering Agriculture. (Available frôm Nottingham Road, Stapleford, Nottingham NG9 8AJ. Tel: 0115 939 0202.)

Among the free leaflets published by the Health and Safety Executive (HSE Information Centre, Broad Lane, Sheffield S3 7HQ. Tel: 0114 289 2345. For free HSE leaflets: HSE Books, P.O. Box 1999, Sudbury, Suffolk, CO10 6FS. Tel: 01787 881165), the following relate to herbicides.

AS 25	*Training in the use of pesticides*
AS 26	*Protective clothing for use with pesticides*
AS 27	*Pesticides on the farm*
AS 28	*COSHH in agriculture*
AS 30	*COSHH in forestry*
CS 19	*Shortage of approved pesticides: guidance for farmers and other professional users.*

Note: Because of the rapid evolution of herbicide practice, readers should ensure that they have the most up-to-date edition of any quoted literature. It should be noted that manufacturers of herbicide products normally reprint their labels annually and may introduce changes at any reprinting.

2 Pesticide legislation

2.1 Introduction

The Food and Environment Protection Act 1985 Part III includes as its aims:

"to protect the health of human beings, creatures and plants; to safeguard the environment and to secure safe, efficient and humane methods of controlling pests"

and

"to make information about pesticides available to the public".

More detailed conditions are laid down in the Control of Pesticides Regulations 1986 (statutory Instrument 1510, 1986).

The Control of Pesticides Regulations 1986 affect those engaged in the application of herbicides.

2.2 Products

Pesticide products are divided into two groups:

a. amateur products:

those pesticide products which are sold to amateur users (i.e. householders) and which can be bought at retail outlets such as garden centres and supermarkets.

b. professional products:

those pesticides which must only be sold to professional users (i.e. those people who are suitably trained to sell, give advice on or use pesticides). These are sold through farm suppliers, and specialist pesticide retailers and wholesalers by certificated staff.

Professional users can use amateur products in the approved manner. But amateurs must not use professional products.

Since October 1986, *all* products marketed as pesticides must have an 'Approval' issued under the regulations. 'Approval' will be given in one of three forms.

Full approval

for an unstipulated period.

A provisional approval

for a stipulated period in order to satisfy any outstanding requirements.

An experimental permit

to enable testing and development to be carried out with a view to providing the Ministers with safety and other data.

2.3 Conditions of approval

Certain conditions of approval must be complied with when using herbicide products. These are set out on the herbicide label and may specify:

- field of use [where the product can be used, within broad categories such as Forestry, Farm Forestry, Woody Ornamentals, Agriculture, Forest Nurseries, Amenity Plantings, etc.]
- crops, plants or surfaces on which the products may be used.
- maximum dose rates
- maximum number of treatments
- maximum harvest intervals
- timing of applications
- statements on operator and environmental protection.

These conditions of approval are legally binding and MUST BE COMPLIED WITH. They are not merely advisory.

2.4 Minor uses

Minor uses are defined by the Pesticides Safety Directorate as "those advantageous uses of pesticide for which anticipated sales volume is not sufficient to persuade the manufacturer to carry out the research and development required to obtain 'full approval' for label recommendations".

As forestry frequently provides a minor use for a herbicide used more widely on other crops, this category of use is of particular interest to foresters. If approved, such minor uses are termed 'off-label approvals'.

Some products with off-label approvals are listed in this book. All users must hold a copy of the relevant off-label approval before using the product. Copies of off-label approvals can be obtained from local ADAS offices (refer to the telephone directory) or from the original applicant.

All applications made under any of the off-label arrangements are at user's own risk. This means that the manufacturers cannot be held responsible for any adverse effects on crops or failure to control weeds, but employers and operators still have the responsibility to adhere to on-label instructions when using the product.

In addition to specific off-label approvals, certain fields of use are covered by long-term off-label arrangements, the revised versions of which are valid until 31 December 1999.

- **All pesticides** with full or provisional label approval for use on any growing crop may be used within forest nurseries, on crops prior to final planting out.
- Christmas trees grown on commercial agricultural and horticultural holdings and in forest nurseries can be regarded as hardy ornamental or forest nursery stock, and are hence covered by the same arrangements.
- **Herbicides** with full or provisional label approval for use on cereals, may be used in the first 5 years of establishment of new farm woodland (including short rotation energy coppice), on land previously under arable cultivation or improved grassland (as defined in the Woodland Grant Scheme).
- **Herbicides** with full or provisional label approval for use on cereals, oilseed rape, sugar beet, potatoes, peas and beans, may be used in the first year of re-growth following cutting in short rotation energy coppice, on land previously under arable cultivation or improved grassland (as defined in the Woodland Grant Scheme).

As well as the usual good working practices required of users, the following **additional** conditions must be complied with when applying pesticides under the long-term off-label arrangements.

- All precautions and statutory conditions of use, which are identified on the product label, must be observed.
- The method of application used must be the same as that listed on the product label, and comply with relevant codes of practice and requirements under COSHH.

- All reasonable precautions must be taken to safeguard wildlife and the environment.
- Products must not be used in or near water unless the label specifically allows such use.
- Aerial applications are not permitted.
- Products approved for use under protection, i.e. under polythene tunnels or glasshouses, cannot be used outside.
- Rodenticides and other vertebrate control agents are not included in these arrangements.
- Use is not permitted on land not intended for cropping, for example paths, roads, around buildings, wild mountainous areas, nature reserves, etc.
- Pesticides classified as hazardous to bees must not be applied when crops or weeds are flowering.
- These extensions of use apply only to label recommendations – no extrapolations are permitted from specific off-label approvals.
- Unless specifically permitted on the product label, handheld applications are **not** permitted if the product label,
  - prohibits handheld use,
  - requires the use of personal protective clothing when using the pesticide at recommended volume rates,
  - is classified as 'corrosive', 'very toxic', 'toxic', or has a 'risk of serious damage to eyes'.

  If none of the above applies, handheld application is permitted provided that,
  - the concentration of the spray volume does not exceed the maximum recommended on the label,
  - spray quality is at least as coarse as the British Crop Protection Council medium or coarse spray (refer to the *Provisional code of practice for the use of pesticides in forestry)*,
  - operators wear a protective coverall, boots and gloves for applications below waist height; a face shield should be worn in addition for applications above waist height,
  - if the product label gives a buffer zone for vehicle mounted use, a buffer zone of 2 m should be used for handheld applications.

Useful addresses:
Forestry Commission
Education, Safety and Training Branch
231 Corstorphine Road
Edinburgh
EH12 7AT

Forestry Commission
Research Division
Silviculture (South) Branch
Alice Holt Lodge
Wrecclesham
Farnham
Surrey
GU10 4LH

Forestry Commission
Research Division
Silviculture (North) Branch
Northern Research Station
Roslin
Midlothian
EH25 9SY

Timber Growers' Association
5 Dublin Street Lane South
Edinburgh
EH1 3PX

Pesticides Safety Directorate
Mallard House, Kings Pool
3 Peasholme Green
York
Y01 2PX

The approved recommendations for major uses of a herbicide will be included on the product label. A useful guide to current on-label approvals can be found in the *UK pesticide guide* and new approvals are notified in *The pesticide register* – see Section 1.9.

2.5 Operators

Under the Control of Pesticides Regulations, general obligations are placed on all those who sell, supply, store or use pesticides.

Certificates of competence are now required by those engaged in the use of pesticides.

A recognised certificate of competence is required by:

- contractors applying approved pesticides unless they are working under the direct and personal supervision of a certificate holder;
- those born on or after 31 December 1964 who are applying approved pesticides unless working under the direct and personal supervision of a certificate holder;
- those required to supervise anyone in the above two categories, who should, but does not hold a current certificate.

Pesticide operations for the purpose of certificate of competence are grouped into 'modules'. Full details are available from:

National Proficiency Tests Council
National Agricultural Centre
Stoneleigh
Warwickshire
CV8 2LC

Contractors who buy and sell pesticides or who advise on the use of pesticides should refer to the *Code of practice for sale or supply of pesticides* issued by MAFF for a description of their liability for:

- competence and certification of any staff who may sell and/or advise on the use of pesticides;
- competence and certification of staff who are responsible for storage of pesticides bought and sold.

Under the Control of Substances Hazardous to Health Regulations 1988, employers are required to make an assessment of the potential risks from using a hazardous substance at work. This considers the risk of exposure to pesticides proposed for use, to employees and others in the vicinity of the operation and identifies appropriate steps to be taken. The employer is then expected to take such steps if he proceeds with the pesticide application. More details on COSHH are given in HSE guidance notes AS27 and AS28 – see Section 1.9.

2.6 Use of adjuvants

The use of adjuvants is now controlled and only those appearing on lists published by the Pesticides Safety Directorate may be used.

If any adjuvant appears on the list but not on the approved product label, it may be used with that product, but at the user's own risk.

Refer to the *UK pesticide guide* for the full list (see Section 1.9).

2.7 Aerial application of pesticides

Aerial applications are now subject to detailed rules, requiring extensive consultations. These are set out in the Control of Pesticides Regulations 1986.

2.8 Codes of practice

A series of codes of practice on pesticides have been published as part of the implementation of part III of the Food and Environment Protection Act 1985.

a. *Code of practice of suppliers of pesticides to agriculture, horticulture and forestry.*

b. *Code of practice for the safe use of pesticides on farms and holdings.*

c. *Code of practice for the safe use of pesticides for non-agricultural purposes.*

d. *Training recommendations for non-agricultural users of pesticides.*

e. *Provisional code of practice for the use of pesticides in forestry,* 1989.

(See Section 1.9).

These codes of practice are intended to help users meet their obligations under the current legislation. Users **must** obtain and refer to the *Code of practice for the safe use of pesticides for non-agricultural purposes.* This forms the definitive source of advice on good working practices, but useful additional information is contained in the *Provisional code of practice for the use of pesticides in forestry*, which it supersedes. The forestry code of practice is divided into guidance notes on the following subjects:

- safe use of pesticides in forestry;

- product approvals;
- competence and skills of users;
- protection of the environment and neighbours' crops;
- safe systems for storage and handling pesticides and associated equipment including stock control and record of usage;
- application equipment;
- reduced volume application of pesticide from ground-based machinery.

In addition the forestry code of practice contains a working check list on the decision to use pesticide, a working check list for operators and a check list of sources of information.

2.9 Poisons Act 1972

Certain products are subject to the provisions of the Poisons Act 1972, the Poisons List Order 1982 and the Poisons Rules 1982 (copies of all these are obtainable from HMSO). These regulations include general and specific provisions for the labelling, storage and sale of scheduled poisons. The scheduled chemicals approved for use in the UK are listed in the *UK pesticide guide* (see Section 1.9). The only herbicide mentioned in this Field Book which is on the poisons list is paraquat.

2.10 The Poisonous Substances in Agriculture Regulations 1984

These regulations have been repealed and their provisions covered by different provisions under the Control of Substances Hazardous to Health Regulations 1988.

3 Safety precautions and good working practices

3

3.1 Private sector training

Operators should check their liability under the The Control of Pesticides Regulations 1986 in relation to the need for certification (see Section 2.5).

Formal training is the most effective way of achieving the level of competence necessary for certification and can be arranged through,

The Forestry and Arboriculture Safety and Training Council
c/o Forestry Commission
231 Corstorphine Road, Edinburgh, EH12 7AT
Telephone: 0131-334 8083.

3.2 The Forestry and Arboriculture Safety and Training Council Safety Guides

The Forestry and Arboriculture Safety and Training Council (FASTCo) has been set up to promote all aspects of safety, particularly safe working practices throughout the forestry industry.

As an aid to maintaining the safe working standards of operators, the FASTCo publishes a series of Safety Guides each of which gives advice on safety in a particular forest operation.

Those currently available relevant to herbicide application are:

Safety Guide 2	*Dipping plants in insecticide*
Safety Guide 4	*Pre-plant spraying of containerised seedlings*
Safety Guide 5	*Application of pesticide by hand held applicators*
Safety Guide 34	*First aid*

These are available from:

The Forestry Authority Safety Officer
Forestry Commission
231 Corstorphine Road
Edinburgh, EH12 7AT

Both the operator and the supervisor should be provided with a copy of the relevant leaflet, which they should read and fully understand before starting the operation.

3.3 Routine precautions

Users planning to apply herbicides must ensure there are proper and safe arrangements from the time of receipt of the container full of herbicide, until the final disposal of the last used container. It is necessary to observe the following general principles to achieve this objective:

- provide for safe storage of herbicides (see Section 3.4);
- comply with label instructions;
- ensure safe transit of herbicides to work site (see Section 3.4);
- wear protective clothing as specified on the label;
- follow label instructions for diluting the concentrated product;
- apply chemicals evenly at correct dose rate;
- avoid any drift on to neighbouring land; particular care is required where applications take place near watercourses – see Section 3.6;
- protect the public from any risk of contamination. Measures that may be necessary include the erection of warning signs [ESSENTIAL if edible fruit is sprayed] and the temporary closure of footpaths.
- minimise any possible harmful impacts on the environment [with regard to wildlife and desirable vegetation for example] by careful choice of herbicides and application methods;
- wash application equipment carefully after use to avoid contamination of crops treated subsequently;
- wash and clean protective equipment regularly;
- dispose of surplus chemical and empty containers safely (see Section 3.9);
- keep adequate records of all operations and staff involved in the application of herbicides.

More detailed guidance on good practice can be obtained from the *Code of practice for the safe use of pesticides for non-agricultural purposes* (see Section 1.9).

3.4 Storage and transport of herbicides

Whenever herbicides are stored, the storage facility should comply with certain criteria.

The store must be:

- suitably sited
- of adequate storage capacity
- soundly constructed of fire resistant materials
- provided with suitable entrances and exits
- capable of containing spillage and leakage
- dry and frost free where necessary
- suitably lit
- suitably ventilated
- marked clearly and secure against vandalism
- equipped and organized to accommodate the intended contents
- equipped with an accurate records system to monitor movements of herbicides in and out of the store.

Full details on the design of storage facilities can be found in the Health and Safety Executive guidance note S19 – see Section 1.9.

When herbicides are transported to the work site, users should check the integrity of product containers, then secure them in a lockable box or chemical safe. The crew compartment of the vehicle must be kept free of contamination from herbicides or protective clothing, and there should be ready supplies of absorbent materials available in case of spillage.

3.5 Herbicides and the environment

Herbicides used sparingly and carefully should have little or no adverse impact on the environment. Indeed, they can be a valuable tool for vegetation management for the benefit of wildlife conservation, for example the removal of invasive species like rhododendron from semi-natural woodlands.

The following principles should be followed to minimise adverse effects on the environment.

- The requirement for herbicide applications should be reduced by timing other establishment operations correctly. Choices may sometimes need to be made between applying herbicides, and other operations such as cultivation. The relative environmental effects of each operation should be compared, but in general the aim should be to minimise the need for herbicides.

- Wherever possible, herbicides that have low toxicity, break down quickly in the soil, and are not washed out of the soil into watercourses or ground water, should be favoured (see Section 3.6). It should be noted that those herbicides that give the longest periods of residual weed control are by definition those that remain active in the soil for the longest periods. Hence, even if there toxicity to animals and insects is low, these products may prevent desirable vegetation, as well as weed species, from re-establishing if they are not targeted correctly.
- A method of application should be chosen that gives the desired level of control with the minimum quantity of active ingredient, and confines activity as far as is practical to the target plants. For example, broad spectrum herbicides such as glyphosate or glufosinate ammonium are likely to damage all vegetation to some degree, even if the plant species are not listed on the product label. If accidental contamination of beneficial ground flora occurs during essential weeding operations, it is quite possible that those species may also be checked or killed. It is often worth considering the use of more selective herbicides with a narrow range of target species. The individual herbicide sections later in this book, together with product labels, give general information on properties of the products. This can be used to make assumptions about effects on non-target species to which reference has not specifically been made. Spot or band treatment is usually sufficient, and is preferable to overall applications.
- The conservation value and sensitivity of the areas to be treated with herbicide should be assessed, and damage to rare or otherwise valuable plant species and communities avoided.
  Refer also to the *Forest nature conservation guidelines* and Forestry Practice Guides 1–8 *The management of semi-natural woodlands* (see Section 1.9).

3.6 Use of herbicides in or near water

3.6.1 Avoiding water contamination

The contamination of water in general, and drinking water in particular, by herbicides **MUST** be avoided. Water undertakers have a statutory duty to limit the concentration of any individual pesticide (including herbicides) in drinking water supplies to less than 0.1 part per billion for individual products, and 0.5 parts per billion in total, regardless of the relative toxicity of different products. Some catchments are especially vulnerable and the likelihood of problems can be established by discussions with the appropriate water authority. There are three principal sources of risk to surface and ground water:

- spillage leading to run-off into streams;
- careless disposal of waste;
- careless application.

Two types of problem may be encountered. Firstly, very small quantities of some herbicides (particularly phenoxy herbicides such as 2,4-D) can create severe taste and odour problems (taint) in drinking water. Secondly, gross pollution can occur as the result of accidental spillage – many herbicides are toxic to fish and other aquatic life at high concentrations.

In the individual herbicide sections later in this book, many products have warnings in the special precautions section regarding their use in or near water. Warnings about harmful effects to fish are also likely to apply to other aquatic life.

The National River Authority in England and Wales, (NRA) and River Purification Boards (RPB) in Scotland should be consulted if large scale use of herbicides is being considered in any surface water catchment area. In upland areas, such bodies should be able to assist in determining the extent of water catchments. It may be more difficult to determine the precise locations and area of catchments of small private supplies for isolated farms and dwellings.

Consultation with water authorities or river purification boards is legally required before aerial application of pesticides on land adjacent to water. Consultations must take place not less than 72 hours before the application commences (see Sections 1.6 and 2.7).

The key points when using herbicides in or near water are as follows:

- Certain herbicides are approved for use in or near water. Foresters needing to apply approved products on the banks and surface water of streams and lakes must rigorously follow the terms of the approvals for such products. Prior agreement of the water authority or river purification board is still required in such circumstances.
- Where a protective strip has been defined alongside a watercourse, only those products approved for use in or near water should be used on such strips.
- Herbicides not approved for use in or near water should not be applied within 10 metres of watercourses and within 20 metres of lakes and reservoirs. A further 10 metres should be allowed when incremental applicators are in use. These distances may need to be increased depending on wind conditions and spray droplet size.
- Care should be taken to minimise unwanted spray drift. This can be achieved (as long as label conditions of use are still adhered to) by selecting larger nozzles that produce larger droplets, using lower spraying pressures, increasing volume rates, lowering spray lance or boom height and through careful consideration of wind speed and direction.
- Streams and lakes must not be used for filling or washing equipment.
- Boreholes, wells and mine shafts must not be used for disposing of waste herbicide, or herbicide containers.
- Residues must not be sprayed on to the ground within 50 metres of any borehole or well.
- The product label should be read, and all conditions of use complied with.
- All chemicals should be stored away from watercourses.
- The overspraying of drains should be avoided, if possible.
- Herbicides should not be applied near watercourses where there is a high risk of run-off – see Section 3.6.2. Risk of run-off may be increased if the ground is frozen, snow covered or baked dry, and when heavy rain is expected.
- A contingency plan should be prepared to deal with accidental spillage. The plan should include relevant contact numbers for

bodies such as the National Rivers Authority or River Purification Board, water companies and downstream landowners. There should also be details of the location and mode of operation of pollution control equipment such as booms and absorbent sheets and pillows. Operators should carry small quantities of these absorbent materials, and have access to further central stocks ready packed in emergency trailers. Refer to Technical Development Branch Report No 7/93 for further details (see Section 1.9).

- In the event of any spillage threatening to enter a stream or lake, the NRA or RPB should be informed immediately.
- More detailed guidance can be found in the *Code of practice for the use of pesticides in forestry*, MAFF Booklet 2078 – *Guidelines for the use of herbicides on weeds in or near watercourses and lakes*, and in the *Forests and water guidelines* (see Section 1.9).

A list of products with approval for use in or near water is given in Section 12.6

3.6.2 Leaching and run-off risks associated with herbicide applications

Even with correct storage, handling, application and disposal of herbicides, water may potentially be polluted, either by leaching through the soil into ground water reserves or, once bound with soil particles, washing away with sediment into surface waters (run-off). The degree of risk is determined by factors such as herbicide physical properties, application rate and method, soil properties and weather patterns.

There are three herbicide properties which influence movement in the soil: soil adsorption, water solubility and persistence. Table 1 summarises information on these properties for approved forestry products. Large values for soil adsorption indicate a herbicide which becomes strongly bound to the soil and hence has a low probability of leaching, but may move with soil particles if erosion takes place. A low value indicates a herbicide which is more likely to move with water and hence leach through the soil. Water solubility is another indication of leaching risk. A herbicide with high solubility will have

Table 1 Run-off and leaching ratings for approved forestry herbicides

Herbicide	Potential run-off rating	Potential leaching rating
Ammonium sulphamate	na	na
Asulam	medium	large
Atrazine	medium	large
2,4-D	medium	large
Clopyralid	large	large
Cyanazine	small	medium
Dalapon	large	large
Dicamba	large	large
Dichlobenil	medium	large
Diquat	large	small
Fluazifop-p-butyl	medium	small
Fosamine ammonium	small	medium
Glufosinate ammonium	small	medium
Glyphosate	large	small
Imazapyr	medium	large
Isoxaben	large	small
Metazachlor	na	na
Paraquat	large	small
Pendimethalin	large	small
Propaquizafop	na	na
Propyzamide	large	large
Triclopyr	medium	medium

na = information not available

greater potential for leaching. Persistence measures the length of time that a herbicide remains in the soil before it is broken down. The longer the half-life, the greater the potential for herbicide movement through run-off or leaching. Table 1 summarises herbicide properties into three classes of run-off and leaching ratings.

Soil also varies in its ability to promote leaching and run-off. Well drained soil types such as podzols are very susceptible to leaching whereas surface water gleys are liable to surface run-off and hence have greater potential for herbicide to be transported to water through being bound to soil particles. Table 2 summarises the relative risk of leaching and run-off for common British soil types under forest.

To assess the risk of herbicide leaching or run-off, Tables 1 and 2 should be consulted in conjunction with one another. Table 3 can then be used to determine potential risk. Risk 1 indicates that the particular herbicide chosen has a high probability of leaching through, (Table 3a) or running off (Table 3b), the soil where application is intended. Users should consider alternative herbicides, particularly in areas with high rainfall. Risk 2 suggests that the

Table 2 **Relative soil vulnerability to run-off and leaching**

Soil type	SCDB* code	Relative run-off risk	Relative leaching risk
Brown earths	10	Low	High
Brown earth	11	Low	High
Podzolic brown earth	12	Low	High
Base rich brown earth	13	Low	High
Podzols	20	Low	High
Typical podzol	21	Low	High
Peaty podzol	22	Mod	High
Ironpan soils	30	High	Low
Intergrade ironpan	31	Mod	Mod
Ironpan soil	32	High	Low
Peaty ironpan soil	33	High	Low
Podzol ironpan with induration	34	Mod	Low
Podzolic ironpan soil	35	Mod	Mod
Gley soils	50	High	Variable
Groundwater gley	53	Mod	High
Valley bottom complex	54	High	High
Surface water gley – clay texture	55	High	Low
Surface water gley – loam texture	56	High	Mod
Surface water gley with duration	57	High	Mod
Flushed peat bogs	60	High	Low
Juncus bog	61	High	Low
Molinia bog – moderate nutrients	62	High	Low
Molinia bog – poor nutrients	63	High	Low
Unflushed peat bogs	70	High	Low
Non-flushed blanket	73	High	Low
Sphagnum bog	74	High	Low
Littoral/skeletal soils	80	Low	High
Well drained littoral soil	81	Low	High
Poorly drained littoral soils	82	Mod	High
Skeletal soils, freely drained	83	Low	High
Skeletal soils, poorly drained	84	High	Mod
Calcareous soils	90	Low	High
Rendzina	93	Low	High
Calcareous brown earth	94	Low	High
Calcareous gley	95	High	Low

*SCDB = Forestry Commission Sub-compartment Database

herbicide may leach or run-off, and additional information on the herbicide, site and application is desirable. Risk 3 indicates a low probability of pollution from the soil.

Where weed control is required on a soil type which has a high relative risk of leaching or run-off, consideration should be given to rejecting herbicides which have a large potential rating in favour of those with smaller ratings, wherever possible. For example, for the control of herbaceous weeds in farm forestry, isoxaben has a large run-off rating. On sites with a large risk of run-off, it may therefore be prudent to choose an alternative such as cyanazine, which has a small run-off rating. Similarly, for bracken control, asulam has a large leaching rating. Its use may be best directed to soils such as ironpans where leaching risk is low, and an alternative such as glyphosate considered on podzols where leaching risk is high.

The tables are for guidance only; the decision on which is the most appropriate herbicide to use in a particular circumstance will also depend on an appreciation of herbicide efficacy, ease of use, toxicity, and health and safety issues. Some herbicides may have a relatively high risk of run-off or leaching, but be far less harmful to the environment than some other products which may have a low risk of movement, but are more toxic. These tables aim to give a forest manager some guidance as to the level of environmental risk that may be involved with different herbicide applications.

Table 3 Determining risk from run-off and leaching

(a) Determining the risk of herbicide leaching

Soil leaching rating	Potential herbicide leaching rating		
	Large	Medium	Small
High	Risk 1	Risk 1	Risk 2
Moderate	Risk 1	Risk 2	Risk 3
Low	Risk 2	Risk 3	Risk 3

(b) Determining the risk of herbicide surface run-off

Soil run-off potential	Herbicide run-off rating		
	Large	Medium	Small
High	Risk 1	Risk 1	Risk 2
Moderate	Risk 1	Risk 2	Risk 3
Low	Risk 2	Risk 3	Risk 3

After Becker *et al.* (1991). A pesticide's risk of leaching or run-off. *American Nurseryman*, April 15, 1991, 108–111.

3.7 Tank mixes

The Control of Pesticides Regulations 1986 states that "…. no person shall combine or mix for use two or more pesticides which are anti-cholinesterase compounds unless the approved label of at least one of the pesticide products states that the mixture may be made, and no person shall combine or mix for use two or more pesticides if all the conditions of the approval relating to this use cannot be complied with".

Hence for products other than anti-cholinesterase compounds (none is listed in this Field Book), tank mixes are permissible as long as all the conditions of use for all the products to be used are complied with. However, unless such mixes are specifically listed on the product label, they are made at the user's own risk.

3.8 Reduced volume applications

Applications of herbicides at rates specified on product labels, but in a smaller volume of diluent than specified in the approval, is permitted in some instances. This may be useful if there is no label recommendation for the preferred application equipment, such as the Microfit Herbi, forestry spot gun, or the Ulvaforest. Where the recommended volume rate is too high for the relevant application equipment, or where applicators are not specifically mentioned on the product label, there may be no reference to them in the herbicide sections later in this book. However, subject to the following conditions, there may be an opportunity to use equipment which delivers volumes lower than those specified on the label.

Reduced-volume application is NOT PERMITTED if:

a. the hazard classification states that the product is 'corrosive', 'very toxic' or 'toxic' or that there is a risk of 'serious damage to eyes';
b. the label states that protective clothing (which includes gloves for adjusting nozzles and handling the spray boom) is required to be worn when applying diluted spray at the label recommended volume rate;
c. reduced-volume spraying is specifically prohibited on the label.

Reduced-volume application IS PERMITTED for other products down to one-tenth of the recommended minimum application volume rate provided that the following protective clothing is worn;

a. for vehicle mounted sprayers (spray equipment carried or trailed by a vehicle and which is operated by a person on the vehicle) – WEAR SUITABLE GLOVES when adjusting nozzles and handling spray boom.

b. for hand held sprayers (spray equipment, self propelled or not, which is operated by a person on foot) – WEAR SUITABLE PROTECTIVE CLOTHING (OVERALLS), GLOVES AND BOOTS.

If faceshield or eye protection is required for handling the concentrate, similar protective equipment must be worn if using 'Reduced-volume' spraying through hand-held equipment.

If tractor spraying operations are conducted without a cab, the protective clothing requirements are the same as those prescribed for the operators of hand-held sprayers.

With reduced volume sprays of less than 250 litres/hectare, vehicle mounted sprays must not be used with VERY FINE spray quality nozzles (<90 μm volume median diameter rotary atomisers). Hand held sprays must not be used with VERY FINE OR FINE spray quality nozzles (<201 μm volume median diameter for rotary atomisers) unless specifically covered by label approvals. Users must, in addition, have instructions on how to control the spray quality of the operation equipment, and keep within all the label conditions of use for the herbicide product.

Users should refer to the *Code of practice for the safe use of pesticides for non-agricultural purposes* for further information and guidance on this subject (see Section 1.9).

3.9 Safe disposal of surplus herbicide

3.9.1 Unused dilute herbicide

In well controlled application, there will be very little unwanted dilute herbicide at the end of a day's work. Small volumes of dilute herbicide should be sprayed on to the previously treated crop area, or on to untreated crops, away from watercourses, avoiding susceptible crop trees, areas with conservation value due to the vegetation or wildlife present, areas with high visitor pressure, and water-saturated ground. The maximum application rate per treated hectare, as stated on the product label, must not be exceeded.

Alternatively, subject to approval from the National Rivers Authority or from River Purification Boards, users may be permitted to spray dilute washings on to non-crop land, provided that it is of minimal wildlife value, supporting only poor vegetation, and with no hedges, trees or bushes on it or nearby. To be approved the land must be capable of absorbing the application without run-off or puddling, or risk to wildlife, watercourses or drainage systems. Where necessary the land must be signed and fenced to exclude people and livestock. The spray tank should be completely drained.

Larger volumes of dilute herbicide, not used because of machine failure, increased wind speed or other reasons, should be returned overnight to a safe store and used as soon as possible thereafter for the original intention. If, however, manufacturers advise that dilute herbicide may denature if kept for more than a day or so, and the herbicide cannot be used within this time, the excess liquid should be sprayed safely on to waste ground away from watercourses as detailed in the preceding paragraph.

3.9.2 Surplus herbicide concentrate

Unopened sound containers of herbicide, surplus to user's requirements, should be offered back to the supplier as soon as it is apparent that a material is surplus. Any surplus material which a supplier will not take back should be disposed of either by prior arrangement with the local authority or by a reputable waste disposal contractor. Under the Environmental Protection Regulations 1992 users are required to obtain a controlled waste transfer note from the Local Authority or disposal contractor, and retain it for 2 years after disposal of surplus material.

3.9.3 Old or deteriorated herbicide

Herbicide should not be kept beyond any date given on a 'Use Before' label or, if there is no such label, for more than 2 years from the date of purchase unless otherwise advised by the manufacturers. Herbicide concentrates showing signs of change (e.g. loss of solvent leading to shrinkage of the container, irreversible settling out, etc.) must not be used. Old or deteriorated herbicide should be disposed of as for surplus herbicides.

3.9.4 Contaminated solid material

Used herbicide concentrate containers should be thoroughly cleaned if possible (without contaminating watercourses or the wider environment), then stored in a sealed plastic bag prior to collection by an disposal contractor – a controlled waste transfer note must be acquired and retained for 2 years after disposal. Dilute washings should be disposed of as per Section 3.9.1.

Solid waste material arising from clean up of chemical spillages, and discarded contaminated protective clothing should again be disposed of by a reputable disposal contractor, and a transfer note retained.

4 Grasses, herbaceous broadleaved weeds and mixtures

4.1 General

Competition by grasses and herbaceous broadleaved weeds in young plantations can seriously reduce the survival and early growth of the crop trees and lead to an extended establishment period. Grasses especially, can compete vigorously for light, nutrients and, in the lowlands and drier uplands, for water. Effective control is therefore usually essential for successful crop establishment and growth.

As forest weeds, grasses can be grouped into two categories: coarse grasses which are generally tall, bulky, rank, stiff, often rhizomatous and tussocky and others which, in contrast, are known as soft grasses. Soft grasses are generally more susceptible to herbicides while coarse grasses usually show a somewhat greater resistance (see Table 4).

The 13 herbicides with full forestry approval or off-label approval that are of use against these weed types are as follows:

Grass control only:

atrazine
propyzamide

Broadleaved herbaceous weed control only:

clopyralid
2,4-D
2,4-D/dicamba/triclopyr
isoxaben
triclopyr

Grass and herbaceous weed mixes:

dalapon/dichlobenil
glufosinate ammonium
glyphosate
imazapyr
paraquat
paraquat/diquat

The effectiveness of these herbicides on the more important grass weeds is given in Table 4. Table 5 lists the herbaceous weed species that have been found to be susceptible to herbicide applications in manufacturers' tests. It is likely that a far wider range of species, and herbaceous vegetation in particular, will be damaged by using these products. This should be borne in mind both when assessing the suitability of individual herbicides for weed control purposes, and when considering their impact on non-target desirable ground cover. For example, a relatively limited range of primarily arable weeds are listed as susceptible to applications of imazapyr. However, it is likely that almost all plant species will be damaged by that herbicide to some degree. Similarly, applications of glyphosate, targeted to control competing bracken or grasses in semi-natural woodland environments, will almost certainly be damaging to desirable woodland herbaceous species if they are accidentally treated. In these circumstances it is worth considering more selective herbicides that control a narrower range of target species. The individual herbicide entries later in this section, and the product labels, can be interpreted to give some indication of likely effects on species to which no specific reference is given.

Perennial rhizomatous grasses are the most difficult grasses to control and require the use of residual herbicides or herbicides which are absorbed by the foliage and translocated to the rhizomes. Residual herbicides can be inactivated by the presence of organic matter in the soil. However, dalapon/dichlobenil, and imazapyr retain some activity on peat soils.

Atrazine is a foliar and soil acting herbicide which is primarily of use against soft grasses. Propyzamide gives soil acting residual control of most grasses, although it is not recommended on peat soils with a peat layer greater than 10 cm in depth. Both atrazine and propyzamide will have some effect on herbaceous weeds, but when these weeds comprise a significant part of the weed population an alternative herbicide should be chosen. Collectively, clopyralid, 2,4-D, 2,4-D/dicamba/triclopyr, isoxaben, and triclopyr will control most broadleaved herbaceous species, but will have little effect on grasses. Isoxaben is a wholly soil acting residual herbicide, which requires a brash and weed-free cultivated planting site to work effectively – it is probably only appropriate for new planting sites. Clopyralid, 2,4-D, 2,4-D/dicamba/triclopyr, and triclopyr are foliar acting herbicides. Where there is a mixture of grasses and herbaceous broadleaved weeds, dalapon/dichlobenil, glufosinate ammonium, glyphosate, or imazapyr should be used. Dalapon/dichlobenil is a mainly soil acting residual herbicide. Imazapyr has foliar and soil action and is translocated. Glufosinate ammonium is foliar acting, and glyphosate is foliar acting and translocated.

The most appropriate herbicide for a given problem will vary depending on the crop present, weed spectrum and season. Figures 1 to 3 and the loose inserts included with this Field Book comprise decision trees which will help in choosing a suitable herbicide.

It is perhaps dangerous to generalise, but in pre-plant situations imazapyr will probably give the broadest spectrum of weed control and prevent weed re-growth the longest. For winter applications propyzamide is very effective against grasses, and dalapon/dichlobenil may give control of herbaceous broadleaved species as well. For spring applications, atrazine will control soft grasses. Glufosinate ammonium will give some control of a wide spectrum of weeds all year round, but glyphosate is translocated so may control deep rooted species better. Neither of these two products provides any long term residual control of germinating weeds. Specific broadleaved herbaceous weeds may be controlled with clopyralid, 2,4-D, 2,4-D/diacamba/triclopyr, isoxaben and triclopyr, with little effect on grasses present. These suggestions are totally dependent on crop species present, and may be further modified by the presence of other weed types such as woody weeds and bracken. Always refer to the specific herbicide sections in the book, and to the product labels.

Paraquat, and paraquat/diquat have similar uses to glufosinate ammonium. Paraquat products are not treated in detail in this book because there are less poisonous alternatives. However, this does not mean that these products are not available for use in forest situations.

Table 4 Susceptibility of common grasses in the forest to approved herbicides

Grass species	Herbicide				
	Atrazine	Glufosinate ammonium	Glyphosate	Propyzamide	Imazapyr
Agropyron repens (couch grass) (C)	MR	MS	S	S	S
Agrostis spp. (Bent grasses)	S	MS	S	S	S
Anthoxanthum odoratum (sweet vernal)	S	MS	S	S	–

Abbreviations:

S = susceptible: control should be excellent.
MS = moderately susceptible: control should be adequate; but grass may rapidly reinvade.
MR = moderately resistant: control may be inadequate.
R = resistant: little effect or control obtained.
(C) = coarse grasses (all others can be considered as soft grasses).
– = information not available.

Notes:

1. No specific information is available on Dalapon/Dichlobenil - refer to individual section.
2. Refer to individual sections and product labels for specific dose rates.

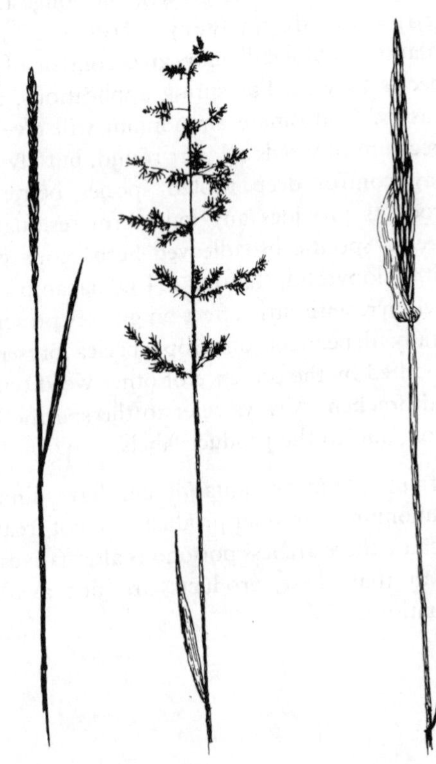

Agropyron repens *Agrostis gigantea* *Anthoxanthum odoratum*

Table 4 Susceptibility of common grasses in the forest to approved herbicides

Grass species	Herbicide				
	Atrazine	Glufosinate ammonium	Glyphosate	Propyzamide	Imazapyr
Arrhenatherum elatius (false oat) (C)	MR	MS	S	S	S
Calamagrostis epigejos (small reed grass) (C)	R	MS	MS	R	S
Dactylis glomerata (cocksfoot) (C)	MR	MS	S	MR	S
Deschampsia caespitosa (tufted hair grass) (C)	MR	MS	S	S	S

Abbreviations:

S = susceptible: control should be excellent.
MS = moderately susceptible: control should be adequate; but grass may rapidly reinvade.
MR = moderately resistant: control may be inadequate.
R = resistant: little effect or control obtained.
(C) = coarse grasses (all others can be considered as soft grasses).
– = information not available.

Notes:

1. No specific information is available on Dalapon/Dichlobenil - refer to individual section.
2. Refer to individual sections and product labels for specific dose rates.

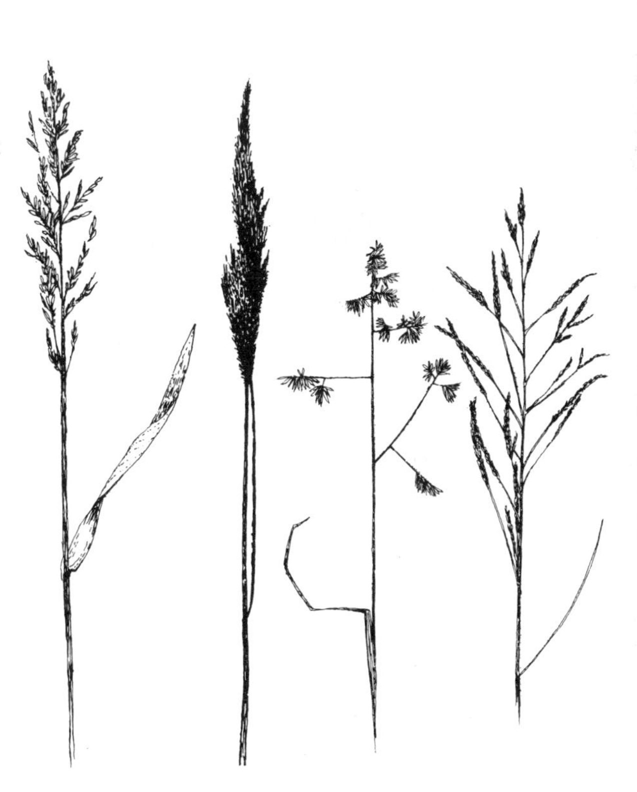

Arrhenatherum elatius *Calamagrostis epigejos* *Dactylis glomerata* *Deschampsia caespitosa*

Table 4 Susceptibility of common grasses in the forest to approved herbicides

Grass species	Herbicide				
	Atrazine	Glufosinate ammonium	Glyphosate	Propyzamide	Imazapyr
Deschampsia flexuosa (wavy hair grass)	S	MS	S	S	S
Festuca arundinacea (tall fescue)	MS	MS	S	S	–
Festuca ovina (sheep's fescue)	S	MS	S	S	–
Festuca pratensis (meadow fescue)	MS	MS	S	S	–

Abbreviations:

S = susceptible: control should be excellent.
MS = moderately susceptible: control should be adequate; but grass may rapidly reinvade.
MR = moderately resistant: control may be inadequate.
R = resistant: little effect or control obtained.
(C) = coarse grasses (all others can be considered as soft grasses).
– = information not available.

Notes:

1. No specific information is available on Dalapon/Dichlobenil - refer to individual section.
2. Refer to individual sections and product labels for specific dose rates.
3. *Festuca pratensis* is very similar in appearance to *Festuca ovina.*

Deschampsia flexuosa *Festuca arundinacea* *Festuca ovina*

Table 4 Susceptibility of common grasses in the forest to approved herbicides

Grass species	Herbicide				
	Atrazine	Glufosinate ammonium	Glyphosate	Propyzamide	Imazapyr
Festuca rubra (red fescue)	S	MS	S	S	–
Holcus lanatus (Yorkshire fog)	S	MS	S	S	S
Holcus mollis (creeping soft grass)	MR	MS	S	MS	S
Molinia caerulea (purple moor grass) (C)	R	MS	S	S	S

Abbreviations:
S = susceptible: control should be excellent.
MS = moderately susceptible: control should be adequate; but grass may rapidly reinvade.
MR = moderately resistant: control may be inadequate.
R = resistant: little effect or control obtained.
(C) = coarse grasses (all others can be considered as soft grasses).
– = information not available.

Notes:
1. No specific information is available on Dalapon/ Dichlobenil - refer to individual section.
2. Refer to individual sections and product labels for specific dose rates.

Table 4 Susceptibility of common grasses in the forest to approved herbicides

Grass species	Herbicide				
	Atrazine	Glufosinate ammonium	Glyphosate	Propyzamide	Imazapyr
Poa annua (annual meadow grass)	S	S	S	S	S
Poa pratensis (smooth meadow grass)	MS	MS	S	S	S
Poa trivialis (rough meadow grass)	S	MS	S	S	–
Juncus spp., (rush)	R	MS	S	R	S

Abbreviations:

S = susceptible: control should be excellent.
MS = moderately susceptible: control should be adequate; but grass may rapidly reinvade.
MR = moderately resistant: control may be inadequate.
R = resistant: little effect or control obtained.
(C) = coarse grasses (all others can be considered as soft grasses).
– = information not available.

Notes:

1. No specific information is available on Dalapon/Dichlobenil - refer to individual section.
2. Refer to individual sections and product labels for specific dose rates.

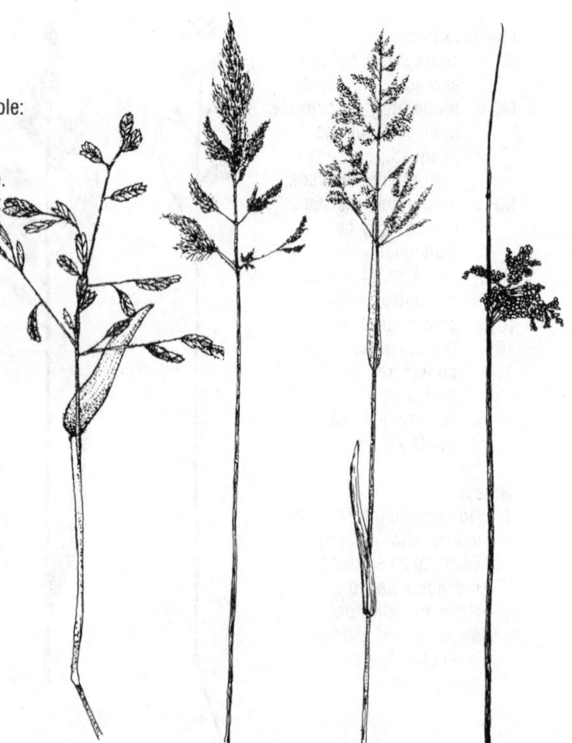

Poa annua *Poa pratensis* *Poa trivialis* *Juncus effusus*

Table 5 Susceptibility of herbaceous vegetation to approved herbicides

Herbaceous weed species	Herbicide										
	Glufosinate ammonium	Imazapyr	2,4-D/dicamba/triclopyr	2,4-D	Clopyralid	Dalapon/dichlobenil	Isoxaben (pre-emergence)	Propyzamide (pre-emergence)	Propyzamide (post-emergence)	Triclopyr	Atrazine (pre-emergence)
Annual mercury											
(Mercurialis annua)	S										S
Annual milk or sow thistle											
(Sonchus asper)	S										
Annual small nettle											
(Urtica urens)	S	S					S	S	MS	S	S
Black bindweed											
(Bilderdykia convolvulus)	S	S			S			S	MS		MS
Black nightshade											
(Solanum nigrum)	S							S	MS		S
Broad dock											
(Ramex obtusifolius)		S	S					S	MS		
Buttercup spp.											
(Ranunculus spp.)			S								
Cat's ear											
(Hypochoeris spp.)				S							
Charlock											
(Sinapis arvensis)	S						S				S

Table 5 Susceptibility of herbaceous vegetation to approved herbicides – *continued*

Herbaceous weed species	Herbicide										
	Glufosinate ammonium	Imazapyr	2,4-D/dicamba/ triclopyr	2,4-D	Clopyralid	Dalapon/ dichlobenil	Isoxaben (pre-emergence)	Propyzamide pre-emergence	Propyzamide (post-emergence)	Triclopyr	Atrazine (pre-emergence)
Cleavers (*Galium aparine*)	S						MS	S	R		MR
Clover spp. (*Trifolium* spp.)				S	S*						
Coltsfoot (*Tussilago farfara*)	S				MS	S					
Common chickweed (*Stellaria media*)	S						S	S	S		S
Common daisy (*Bellis* spp.)			MS	S							
Common dandelion (*Taraxacum officinalis*)	S	S	MS	S		S					
Common orache (*Atriplex patula*)	S						S				
Common vetch (*Vicia sativa*)		S									
Corn buttercup (*Ranunculus arvensis*)	S		S				S				R

Table 5 Susceptibility of herbaceous vegetation to approved herbicides – *continued*

Herbaceous weed species	Herbicide										
	Glufosinate ammonium	Imazapyr	2,4-D/dicamba/ triclopyr	2,4-D	Clopyralid	Dalapon/ dichlobenil	Isoxaben (pre-emergence)	Propyzamide (pre-emergence)	Propyzamide (post-emergence)	Triclopyr	Atrazine (pre-emergence)
Corn chamomile *(Anthemis arvensis)*							S				S
Corn marigold *(Chrysanthemum segetum)*	S				S		S				S
Corn mint *(Mentha arvensis)*	S										
Corn poppy *(Papaver rhoeas)*	S						S				S
Corn spurrey *(Spergula arvensis)*	S						S	S	S		MS
Cow parsley *(Anthriscus sylvestris)*	S										
Crane's bill *(Geranium* spp.)	S	S	S								
Creeping buttercup *(Ranunculus repens)*	S	S	S	S				S	MS		
Creeping cinquefoil *(Potentilla repens)*		S									

Table 5 Susceptibility of herbaceous vegetation to approved herbicides – *continued*

Herbaceous weed species	Herbicide										
	Glufosinate ammonium	Imazapyr	2,4-D/dicamba/ triclopyr	2,4-D	Clopyralid	Dalapon/ dichlobenil	Isoxaben (pre-emergence)	Propyzamide (pre-emergence)	Propyzamide (post-emergence)	Triclopyr	Atrazine (pre-emergence)
Creeping thistle (*Cirsium arvense)*	S	S	S	S	S	S					
Curled dock (*Rumex Crispus)*	S		S			S		S	MS		
Dock spp. (*Rumex* spp.)			S			S					
Fat hen (*Chenopodium album)*	S						S	S	MS		S
Field bindweed (*Convolvulus arvensis)*	S	S									
Field horsetail (*Equisetum arvense)*	S					S		S	MS		
Field pansy (*Viola arvensis)*		S					S				MS
Field penny-cress (*Thlaspi arvense)*	S										S
Field speedwell (*Veronica persicae)*	S										MS

Table 5 Susceptibility of herbaceous vegetation to approved herbicides – *continued*

Herbaceous weed species	Herbicide										
	Glufosinate ammonium	Imazapyr	2,4-D/dicamba/ triclopyr	2,4-D	Clopyralid	Dalapon/ dichlobenil	Isoxaben (pre-emergence)	Propyzamide (pre-emergence)	Propyzamide (post-emergence)	Triclopyr	Atrazine (pre-emergence)
Forget-me-not *(Myosotis arvensis)*	S						S	S	S		S
Fumitory *(Fumaria officinalis)*	S						S	MS	R		MS
Gallant soldier *(Galinsoga parviflora)*	S										
Golden rod *(Solidago canadensis)*		S									
Greater plantain *(Plantago major)*	S	S	S	S		S					
Ground elder *(Aegopodium podagraria)*						S					
Groundsel *(Senecio vulgaris)*	S	S			S		S				S
Hairy bittercress *(Cardamine hirsuta)*							S				
Hawkbit *(Leontodon autumnalis)*				S							

Table 5 Susceptibility of herbaceous vegetation to approved herbicides – *continued*

Herbaceous weed species	Herbicide										
	Glufosinate ammonium	Imazapyr	2,4-D/dicamba/triclopyr	2,4-D	Clopyralid	Dalapon/dichlobenil	Isoxaben (pre-emergence)	Propyzamide (pre-emergence)	Propyzamide (post-emergence)	Trichopyr	Atrazine (pre-emergence)
Hawkweed spp. (*Hieracium* spp.)			S	S							
Hemp nettle (*Galeopsis tetranit*)	S						S				S
Henbit deadnettle (*Lamium amplexicaule*)	S										S
Hoary cress (*Cardarla draba*)	S										
Hoary plantain (*Plantago media*)			S	S		S					
Hogweed (*Heracleum sphondylium*)		S									
Ivy-leaved speedwell (*Veronica hederifolia*)	S						S	S	MS		
Knotgrass (*Polygonum aviculare*)	S						S	S	MS		MR
Meadow buttercup (*Ranunculus acris*)	S		S								

Table 5 Susceptibility of herbaceous vegetation to approved herbicides – *continued*

Herbaceous weed species	Herbicide										
	Glufosinate ammonium	Imazapyr	2,4-D/dicamba/ triclopyr	2,4-D	Clopyralid	Dalapon/ dichlobenil	Isoxaben (pre-emergence)	Propyzamide (pre-emergence)	Propyzamide (post-emergence)	Triclopyr	Atrazine (pre-emergence)
Mouse ear (*Cerastium* spp.)				S							
Mugwort (*Artemisia vulgaris*)		S									
Pale persicaria (*Polygonum lapathifolium*)	S										MS
Parsley piert (*Aphanes arvensis*)	S						S				S
Pearlwort (*Sagina* spp.)				S							
Perennial sow-thistle (*Sonchus arvensis*)	S				S						
Perennial/stinging/common nettle (*Urtica dioica*)	S	S	S			S				S	
Pineapple weed (*Chamomilla suaveolens*)	S				S		S				S
Plantain spp. (*Plantago* spp.)			S			S					S

Table 5 Susceptibility of herbaceous vegetation to approved herbicides – *continued*

Herbaceous weed species	Herbicide										
	Glufosinate ammonium	Imazapyr	2,4-D/dicamba/ triclopyr	2,4-D	Clopyralid	Dalapon/ dichlobenil	Isoxaben (pre-emergence)	Propyzamide (pre-emergence)	Propyzamide (post-emergence)	Triclopyr	Atrazine (pre-emergence)
Ragwort *(Senecio jacobaea)*		S	MS			S					
Red deadnettle *(Lamium purpureum)*	S						S				S
Redshank *(Polygonum persicaria)*	S						S	S	MS		MS
Ribwort plantain *(Plantago lanceolata)*		S	S	S		S					
Scarlet pimpernel *(Anagallis arvensis)*	S						S				S
Scented mayweed *(Chamomilla recutita)*	S				S		S				S
Scentless mayweed *(Matricaria perforata)*	S				S		S				S
Self heal *(Prunella vulgaris)*				S							
Shepherd's purse *(Capsella bursa-pastoris)*	S	S					S	MS	R		S

Table 5 Susceptibility of herbaceous vegetation to approved herbicides – *continued*

Herbaceous weed species	Herbicide										
	Glufosinate ammonium	Imazapyr	2,4-D/dicamba/ triclopyr	2,4-D	Clopyralid	Dalapon/ dichlobenil	Isoxaben (pre-emergence)	Propyzamide (pre-emergence)	Propyzamide (post-emergence)	Triclopyr	Atrazine (pre-emergence)
Smooth hawksbeard *(Crepis capillaris)*	S										
Smooth sow-thistle *(Sonchus oleraceus)*	S	S			S						S
Sorrel spp. *(Rumex* spp.)			S					S	MS		MR
Spear thistle *(Cirsium vulgare)*			S		S						
Spurge *(Euphorbia* spp.)	S										
Stinking chamomile *(Anthemis* spp.)	S				S		S				S
St John's Wort *(Hypercium perforatum)*		S									
Tare spp. *(Vicia* spp.)	S										
Thale cress *(Arabidopsis thaliana)*							S				

Table 5 Susceptibility of herbaceous vegetation to approved herbicides – *continued*

Herbaceous weed species	Herbicide										
	Glufosinate ammonium	Imazapyr	2,4-D/dicamba/triclopyr	2,4-D	Clopyralid	Dalapon/dichlobenil	Isoxaben (pre-emergence)	Propyzamide (pre-emergence)	Propyzamide (post-emergence)	Triclopyr	Atrazine (pre-emergence)
Trefoiles *(Trifolium)*				S	S*						
Vetches *(Vicia* spp.)					S*						
Viper's bugle (Ajuga spp.)						S					
Volunteer oilseed rape *(Brassica napus)*	S						S				
White clover *(Trifolium repens)*		S		S							
Wild carrot *(Paucus carota)*					S						
Wild pansy *(Viola tricolor)*	S										MR
Wild radish *(Raphanus raphanistrum)*	S						S				S

Table 5 Susceptibility of herbaceous vegetation to approved herbicides – *continued*

Herbaceous weed species	Herbicide										
	Glufosinate ammonium	Imazapyr	2,4-D/dicamba/triclopyr	2,4-D	Clopyralid	Dalapon/dichlobenil	Isoxaben (pre-emergence)	Propyzamide (pre-emergence)	Propyzamide (post-emergence)	Triclopyr	Atrazine (pre-emergence)
Willowherb *(Chamerion angustifolium)*	S	S				S		R	R	S	R
Yarrow *(Achillea millefolium)*	S	S		S	MS						

Key:

* - From seed

S - Susceptible

MS - Moderately susceptible

MR - Moderately resistant

R - Reistant

All entries are for POST-EMERGENCE activity on weeds, except where noted for isoxaben, propyzamide and atrazine.

Note: This Table lists only those weed species that have been identified on product labels. Glyphosate is not included in this Table as no specific species are listed – it will give a degree of control over all grass and herbaceous vegetation, except for field horsetail.

It is likely that a wider range of species than those listed in the Table will be controlled, but they have not been the subject of formal experimentation. No guarantees will be made by manufacturers, but it is possible that the following will hold true, in addition to the information listed in the Table.

- Glufosinate ammonium will control most annual grass and herbaceous weeds, but perennial or deep-rooted species may require repeat applications for a complete kill.
- Imazapyr will damage all grasses and herbaceous species to some degree.
- 2,4-D/dicamba/triclopyr, 2,4-D, and triclopyr, will all damage most herbaceous vegetation.
- Dalapon/dichlobenil will control all annual weeds, and control or check most perennial grasses.
- Atrazine will also control the herbaceous weeds listed in the Table as susceptible pre-emergence, if applied in early post-emergence situations when vegetation is less than 50 mm tall.

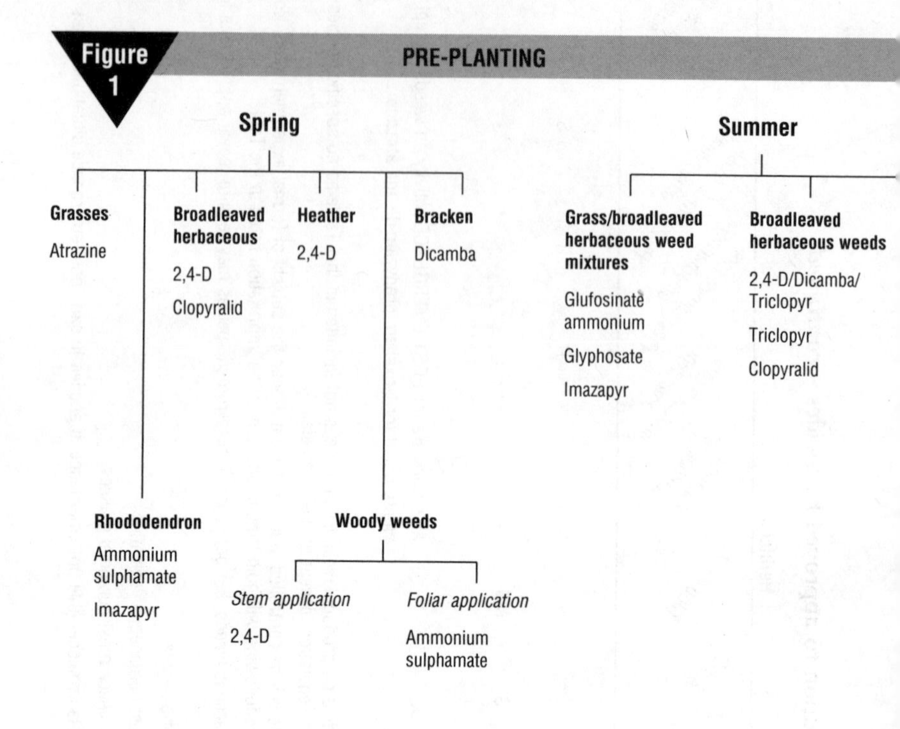

Notes
Decide on the season intended for weed control, then follow the flow diagram by species to identify the potential herbicides.
The diagram should only be used as an aid to identify candidate herbicides – always refer to the relevant sections in the text, and to the product labels, before commencing a weed control programme.

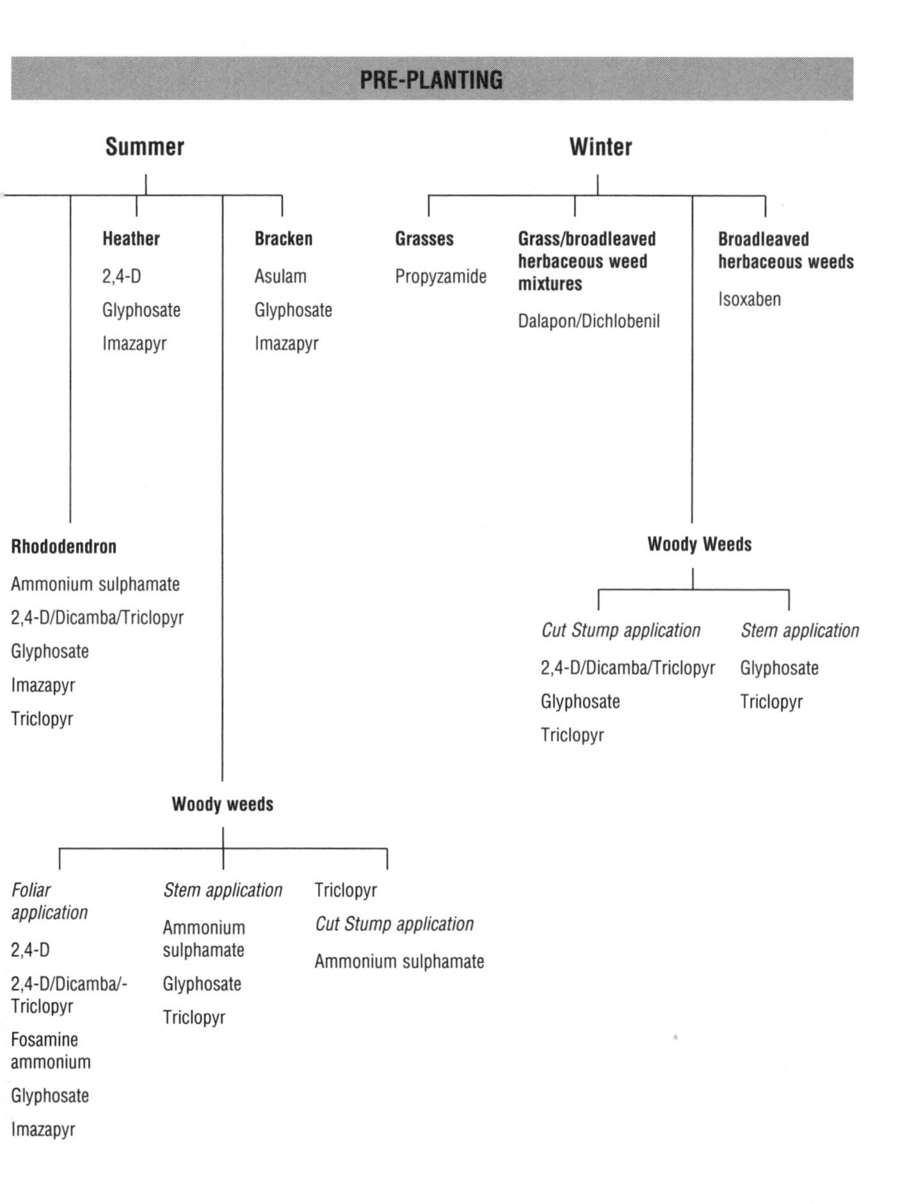
PRE-PLANTING
Summer
Winter
Heather
2,4-D
Glyphosate
Imazapyr
Bracken
Asulam
Glyphosate
Imazapyr
Grasses
Propyzamide
Grass/broadleaved herbaceous weed mixtures
Dalapon/Dichlobenil
Broadleaved herbaceous weeds
Isoxaben
Rhododendron
Ammonium sulphamate
2,4-D/Dicamba/Triclopyr
Glyphosate
Imazapyr
Triclopyr
Woody Weeds
Cut Stump application
2,4-D/Dicamba/Triclopyr
Glyphosate
Triclopyr
Stem application
Glyphosate
Triclopyr
Woody weeds
Foliar application
2,4-D
2,4-D/Dicamba/-Triclopyr
Fosamine ammonium
Glyphosate
Imazapyr
Stem application
Ammonium sulphamate
Glyphosate
Triclopyr
Triclopyr
Cut Stump application
Ammonium sulphamate

4

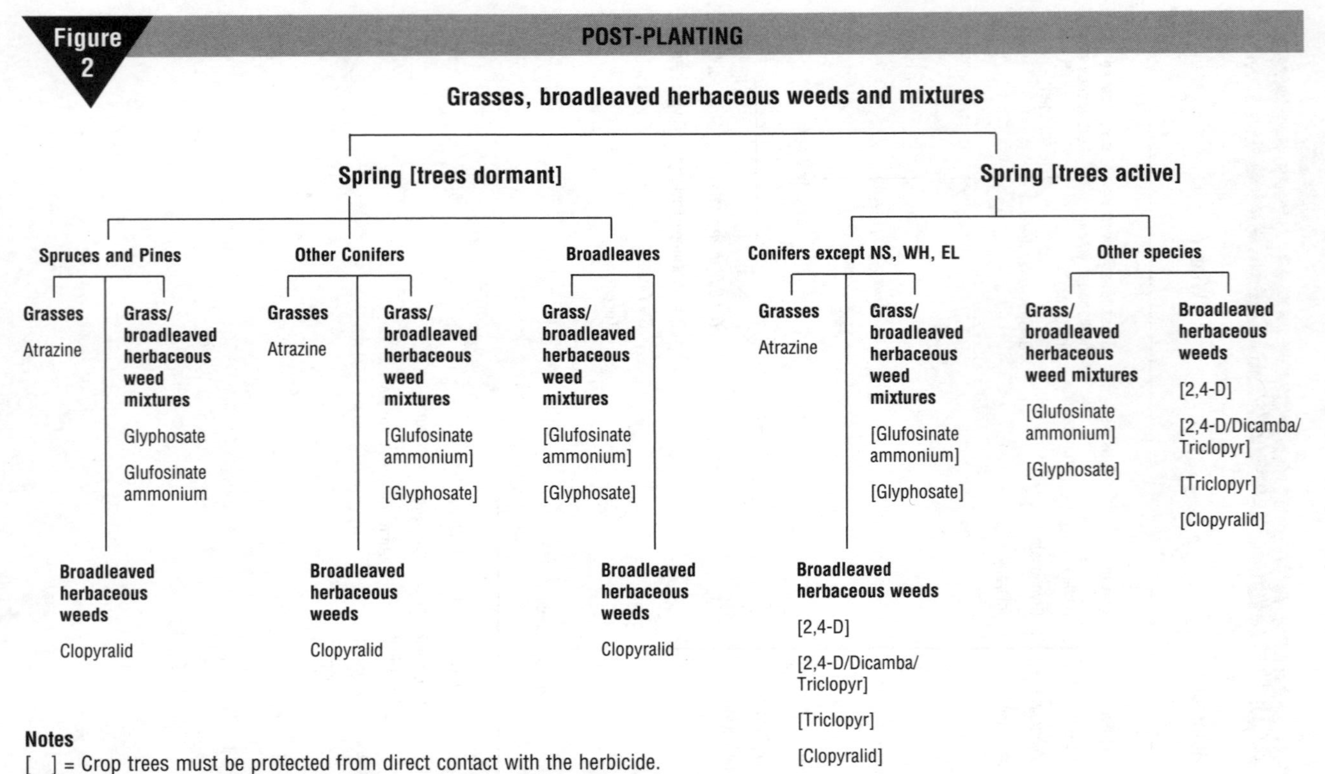
Figure 2
POST-PLANTING
Grasses, broadleaved herbaceous weeds and mixtures
Spring [trees dormant]
Spring [trees active]
Spruces and Pines
Grasses
Atrazine
Grass/ broadleaved herbaceous weed mixtures
Glyphosate
Glufosinate ammonium
Broadleaved herbaceous weeds
Clopyralid
Other Conifers
Grasses
Atrazine
Grass/ broadleaved herbaceous weed mixtures
[Glufosinate ammonium]
[Glyphosate]
Broadleaved herbaceous weeds
Clopyralid
Broadleaves
Grass/ broadleaved herbaceous weed mixtures
[Glufosinate ammonium]
[Glyphosate]
Broadleaved herbaceous weeds
Clopyralid
Conifers except NS, WH, EL
Grasses
Atrazine
Grass/ broadleaved herbaceous weed mixtures
[Glufosinate ammonium]
[Glyphosate]
Broadleaved herbaceous weeds
[2,4-D]
[2,4-D/Dicamba/ Triclopyr]
[Triclopyr]
[Clopyralid]
Other species
Grass/ broadleaved herbaceous weed mixtures
[Glufosinate ammonium]
[Glyphosate]
Broadleaved herbaceous weeds
[2,4-D]
[2,4-D/Dicamba/ Triclopyr]
[Triclopyr]
[Clopyralid]
Notes
[ ] = Crop trees must be protected from direct contact with the herbicide.
Decide on the season intended for weed control, then follow the flow diagram by species to identify the potential herbicides.
The diagram should only be used as an aid to identify candidate herbicides – always refer to the relevant sections in the text, and to the product labels, before commencing

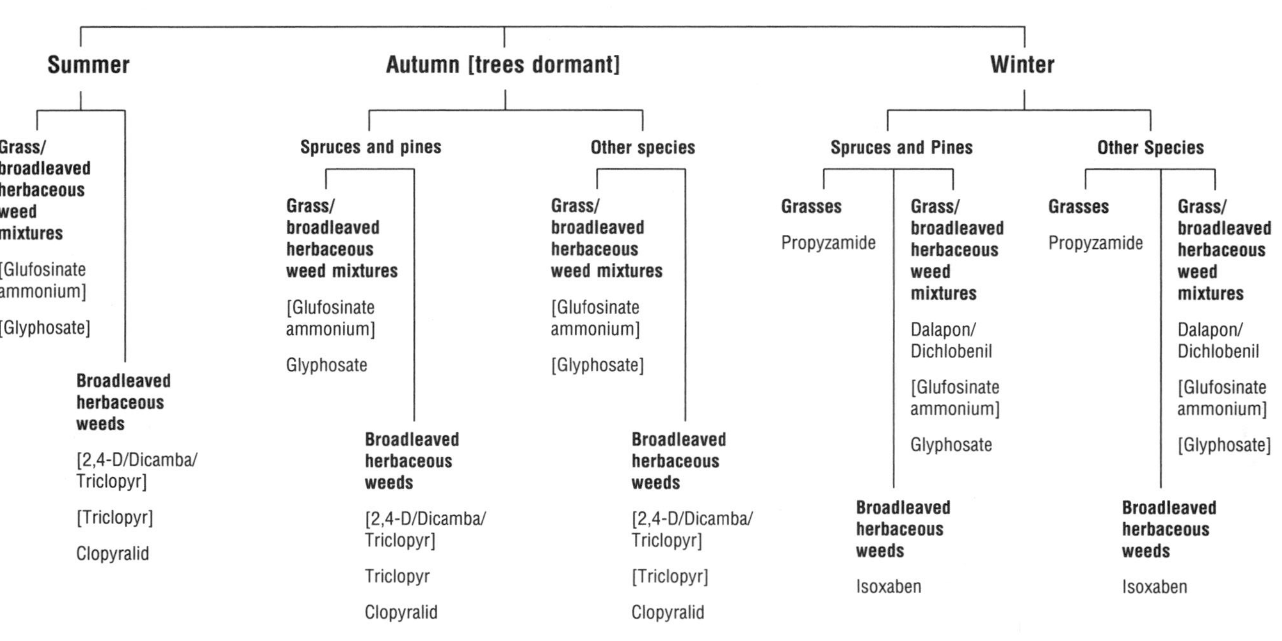

Notes

[ ] = Crop trees must be protected from direct contact with the herbicide.

Decide on the season intended for weed control, then follow the flow diagram by species to identify the potential herbicides.

The diagram should only be used as an aid to identify candidate herbicides – always refer to the relevant sections in the text, and to the product labels, before commencing a weed control programme.

Figure 3 POST PLANTING

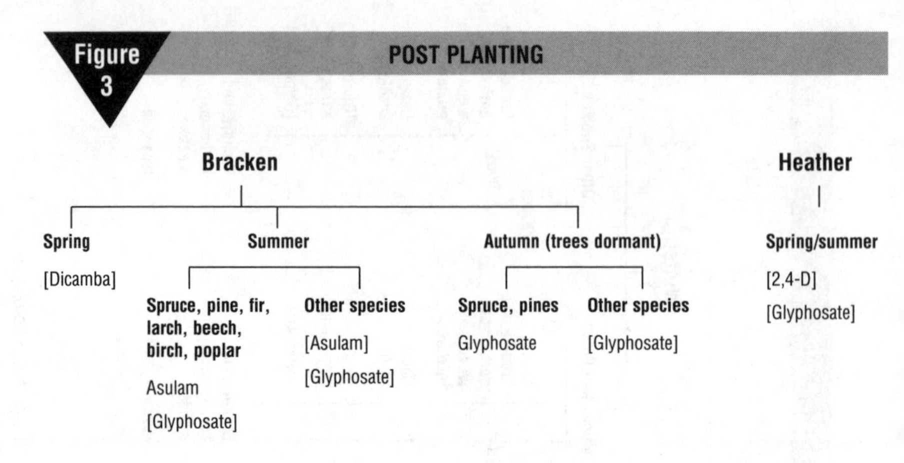

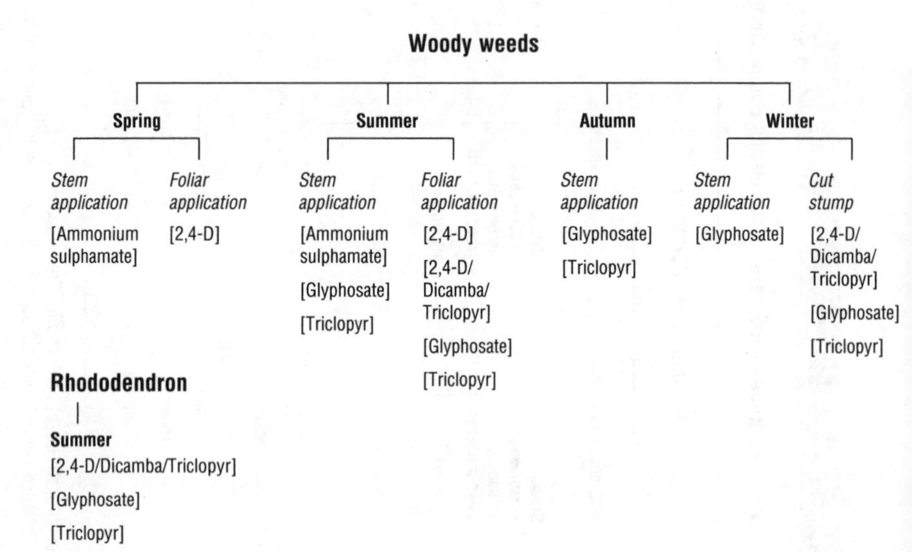

Notes

[ ]= Crop trees must be protected from direct contact with the herbicide.

Decide on the season intended for weed control, then follow the flow diagram by species to identify the potential herbicides.

The diagram should only be used as an aid to identify candidate herbicides – always refer to the relevant sections in the text, and to the product labels, before commencing a weed control programme.

4.2 Atrazine

Product

Unicrop Flowable Atrazine   500g/litre atrazine (Unicrop)

Description

Atrazine has both foliar contact and residual soil action. It is most useful on soft grasses (see Table 1) but will give some control of broadleaved species and the coarse grasses.

Atrazine has no soil activity in soils with an organic peat layer.

Planting can be carried out immediately after treatment but the level of weed control will be reduced if the soil is badly disturbed.

Crop tolerance

Conifers – all the major forest species are tolerant to overall application except NS, WH and EL which are sensitive during the growing season and should only be treated before the start of bud burst, at the lower application rate.

All broadleaves are sensitive while in leaf and should only be treated before the start of bud burst in the spring.

No label approval for use on broadleaves exists, but an off-label approval has been obtained – a copy is included at the end of this Section.

Product rate

Apply 9–13.5 litres of product per treated hectare.

Methods of application

Pre-plant (overall) or post-plant (directed, overall, band or spot application)

Tractor mounted equipment

Boom sprayer at MV.
Ulvaforest low speed rotary atomiser at VLV.

Handheld equipment

Knapsack sprayer at MV.
Knapsack sprayer at LV.
Herbi low speed rotary atomiser at VLV.
Forestry spot gun at LV.

Refer to Section 11 for details of applicators and correct calibration.

Timing of application

February to May one application per year. February and March applications usually give the best results. Applications in May and

June will give effective weed control but also slight crop damage (and severe damage to the more sensitive species NS, WH and larch: see above).

Additional information

1. **Weed control**

a. Atrazine has high solubility, it is therefore unsuitable for use on very light soils, soils with poor structure, i.e. man-made sites, recently constructed sites (which are particularly prone to frost or drought cracking) or sites prone to waterlogging.

   Crops on light calcareous or sandy soils seem to suffer relatively more through damage from atrazine than crops on heavier textured soils.

b. Do not apply atrazine to unhealthy or badly planted trees.

c. Parts a and b above have particular relevance when applications are made around broadleaved species. If there is any doubt over soil type or tree health, it may be safer to aim for rates of 9 litres per hectare or lower. In farm forestry situations a tank mix of 5 litres per hectare of atrazine with 4 litres per hectare of cyanazine (comprising the equivalent rates of active ingredient per hectare as the old Holtox product) may be a more suitable approach. Such a mix is NOT permissible in normal forestry situations.

d. Care must be taken when atrazine is used on steep slopes; heavy rain soon after application can cause creep and surface run-off before the chemical has penetrated the soil.

e. Atrazine does not give an acceptable level of weed control on soils with an organic peat layer, where weeds exceed 3 cm in height.

f. *Holcus mollis,* although a soft grass, is not readily controlled.

g. Do not use atrazine on NS intended for Christmas trees.

h. Users are advised to plant grass strips, or leave strips 10–20 m wide between treated areas and surface waters (see Section 3.6).

2. **Protective clothing**

   Read the product label for protective clothing and equipment requirements and check there are no items required in addition to the Forestry Commission recommendations in Section 10.

3. **Special precautions**

   Atrazine concentrate can be harmful to fish; do not contaminate ponds, watercourses or ditches with the chemical or used containers.

The label on the herbicide container has been designed for your protection – ALWAYS READ THE INSTRUCTIONS ON THE LABEL.

Additional atrazine product with full forestry approval

Atlas Atrazine	500g/litre atrazine (Atlas)

This product may have been different conditions of use – refer to individual labels.

4

NOTICE OF APPROVAL No. 1338/94

FOOD AND ENVIRONMENT PROTECTION ACT 1985
CONTROL OF PESTICIDES REGULATIONS 1986
(S.I. 1986 No. 1510):
APPROVAL FOR OFF-LABEL USE OF AN APPROVED PESTICIDE PRODUCT

This approval provides for the use of the product named below in respect of crops and situations, other than those included on the product label. Such 'off-label use', as it is known, is at all times done at the user's choosing, and the commercial risk is entirely his or hers.

The conditions below are statutory. They must be complied with when the off-label use occurs. Failure to abide by the conditions of approval may constitute a breach of that approval, and a contravention of the Control of Pesticides Regulations 1986. The conditions shown below supersede any on the label *which would otherwise apply.*

Level and scope:	In exercise of the powers conferred by regulation 5 of the Control of Pesticides Regulations 1986 (SI 1986/1510) and of all other powers enabling them in that behalf, the Minister of Agriculture, Fisheries and Food and the Secretary of State, hereby jointly give full approval for the use of
Product name:	Unicrop Flowable Atrazine containing
Active ingredient:	500 g/l atrazine
Marketed by:	Universal Crop Protection Ltd under MAFF No. 02268 subject to the conditions relating to off-label use set out below:

Date of issue:	16 August 1994
Date of expiry:	Unlimited, subject to the continuing approval of MAFF 02268
Field of use:	ONLY AS A FORESTRY HERBICIDE
Crop/situation:	Forestry (broadleaf)
Maximum individual dose:	13.5 litres product/hectare
Maximum number of treatments:	One per year
Operator protection:	*(1)* Engineering control of operator exposure must be used where reasonably practicable in addition to the following personal protective equipment:

(a) Operators must wear suitable protective clothing (coveralls) and suitable gloves when handling contaminated surfaces.

(b) Operators must wear suitable protective clothing (coveralls) when applying by vehicle-mounted equipment.

(c) Operators must wear suitable protective clothing (coveralls), suitable protective gloves, boots, face protection (faceshield) and respiratory/ protective equipment (disposable filtering facepiece respirator) when applying by hand-held equipment.

(d) Operators must wear suitable protective clothing (coveralls), suitable protective gloves and face protection (faceshield) when handling the concentrate.

(2) However, engineering controls may replace personal protective equipment if a COSHH assessment shows they provide an equal or higher standard of protection.

Environmental protection:

(1) Since this product is dangerous to fish or aquatic life and acquatic higher plants, surface waters or ditches must not be contaminated with chemical or used container.

(2) Since there is a risk to aquatic life from use, direct spray from ground-based vehicle-mounted/drawn sprayers must not be allowed to fall within 6 m of surface waters or ditches; direct spray from handheld sprayers must not be allowed to fall within 2 m of surface waters or ditches; spray must be directed away from water.

Other specific restrictions:

(1) This product must only be applied if the terms of this approval, the product label

and/or leaflet and any additional guidance on off-label approvals have first been read and understood.

(2) Use must be restricted to one product containing atrazine or simazine either to a single application at the maximum approved rate or (subject to any existing maximum permitted number of treatments) to several applications at lower doses up to the maximum approved rate for a single application.

4

Signed J Micklewright
(Authorised signatory)

Date 16 August 1994

Application Reference Number: COP 87/00873

THIS NOTICE OF APPROVAL IS NUMBER 1338 of 1994

ADVISORY INFORMATION

This approval relates to the use of 'Unicrop Flowable Atrazine' as a herbicide for use in broad-leaved forestry or woodland, to control grass and herbaceous weeds. The product is applied from February to May, in a minimum of 200 litres water/hectare, by means of tractor-mounted or handheld sprayers.

Please note that an identical off-label approval has been issued for MAFF 05446.

NOTICE OF APPROVAL No. 1339/94

FOOD AND ENVIRONMENT PROTECTION ACT 1985
CONTROL OF PESTICIDES REGULATIONS 1986
(S.I. 1986 No. 1510):
APPROVAL FOR OFF-LABEL USE OF AN APPROVED PESTICIDE PRODUCT

This approval provides for the use of the product named below in respect of crops and situations, other than those included on the product label. Such 'off-label use', as it is known, is at all times done at the user's choosing, and the commercial risk is entirely his or hers.

The conditions below are statutory. They must be complied with when the off-label use occurs. Failure to abide by the conditions of approval may constitute a breach of that approval, and a contravention of the Control of Pesticides Regulations 1986. The conditions shown below supersede any on the label *which would otherwise apply*.

Level and scope:	In exercise of the powers conferred by regulation 5 of the Control of Pesticides Regulations 1986 (SI 1986/1510) and of all other powers enabling them in that behalf, the Minister of Agriculture, Fisheries and Food and the Secretary of State, hereby jointly give full approval for the use of
Product name:	Unicrop Flowable Atrazine containing
Active ingredient:	500 g/l atrazine
Marketed by:	Universal Crop Protection Ltd under MAFF No. 05446 subject to the conditions relating to off-label use set out below:

Date of issue:	16 August 1994
Date of expiry:	Unlimited, subject to the continuing approval of MAFF 05446
Field of use:	ONLY AS A FORESTRY HERBICIDE
Crop/situation:	Forestry (broadleaf)
Maximum individual dose:	13.5 litres product/hectare
Maximum number of treatments:	One per year
Operator protection:	*(1)* Engineering control of operator exposure must be used where reasonably practicable in addition to the following personal protective equipment: (a) Operators must wear suitable protective clothing (coveralls) and suitable protective gloves when handling contaminated surfaces. (b) Operators must wear suitable protective clothing (coveralls) when applying by vehicle-mounted equipment. (c) Operators must wear suitable protective clothing (coveralls), suitable protective gloves, boots, face protection (faceshield) and respiratory protective equipment (disposable filtering facepiece respirator) when applying by hand-held equipment.

4

(d) Operators must wear suitable protective clothing (coveralls), suitable protective gloves and face protection (faceshield) when handling the concentrate.

(2) However, engineering controls may replace personal protective equipment if a COSHH assessment shows they provide an equal or higher standard of protection.

Environmental protection:

(1) Since this product is dangerous to fish or aquatic life and acquatic higher plants, surface waters or ditches must not be contaminated with chemical or used container.

(2) Since there is a risk to aquatic life from use, direct spray from ground-based vehicle-mounted/drawn sprayers must not be allowed to fall within 6 m of surface waters or ditches; direct spray from handheld sprayers must not be allowed to fall within 2 m of surface waters or ditches; spray must be directed away from water.

Other specific restrictions:

(1) This product must only be applied if the terms of this approval, the product label

and/or leaflet and any additional guidance on off-label approvals have first been read and understood.

(2) Use must be restricted to one product containing atrazine or simazine either to a single application at the maximum approved rate or (subject to any existing maximum permitted number of treatments) to several applications at lower doses up to the maximum approved rate for a single application.

Signed J Micklewright
(Authorised signatory)

Date 16 August 1994

Application Reference Number: COP 87/00873

THIS NOTICE OF APPROVAL IS NUMBER 1339 of 1994

ADVISORY INFORMATION

This approval relates to the use of 'Unicrop Flowable Atrazine' as a herbicide for use in broad-leaved forestry or woodland, to control grass and herbaceous weeds. The product is applied from February to May, in a minimum of 200 litres water/hectare, by means of tractor-mounted or handheld sprayers.

Please note that an identical off-label approval has been issued for MAFF 02268.

4.3 Clopyralid

Product

Dow Shield 200 g/litre clopyralid (DowElanco)

Description

A foliar acting herbicide for the control of some annual and perennial dicotyledons. This product has off-label approval for use in conifers and broadleaved trees (applied for by TGA) – all applications are at the user's own risk. The manufacturers hold no responsibility for any adverse effects on crops, or failure to control weeds. A copy of the off-label approval is given at the end of this section. However, employers and operators must still comply with on-label recommendations. Detail on specific weeds controlled by this product is given in Table 7, Section 9.

Crop tolerance

In small scale Forestry Commission trials, it was found that SS, NS, DF, NF, CP, SP, WRC, JL, oak, ash, sycamore, beech, wild cherry, birch, alder, sweet chestnut, Norway maple, poplar and willow will tolerate overspraying by clopyralid, at the rate given, with little resultant damage. After flushing, trees may tolerate overspray as long as young growth has hardened off, but directed sprays should be used wherever possible.

Product rate

Apply 0.5–1 litres of product per treated hectare.

Methods of application

Pre- or post-plant (overall band, spot or directed spray)

Tractor mounted equipment

Boom sprayer at MV.

Handheld equipment

Knapsack sprayer at MV.

Forestry spot gun at LV.

Avoid run-off from treated leaves.

Refer to Section 11 for details of applicators and correct calculation.

Timing

Timing of application will be dependent on growth stage of the target weeds – see Table 4, Section 9. In general, best results will be obtained from applications to young, actively growing weeds. Up to two applications a year may be made.

Additional information

1. **Weed control**

a. Do not apply if rainfall is expected within 6 hours.

2. **Protective clothing**

Read the product label for protective clothing and equipment requirements, and check there are no items required in addition to the Forestry Commission recommendations in Section 10.

3. **Special precautions**

a. Livestock should be kept out of treated areas for at least 7 days after application, and until the foliage of any poisonous weeds such as ragwort has died down and become unpalatable.

b. Do not contaminate ponds, watercourses or ditches with the chemical or the used container.

c. The label on the herbicide container has been designed for your protection – ALWAYS READ THE INSTRUCTIONS ON THE LABEL.

NOTICE OF APPROVAL No. 0757/92

FOOD AND ENVIRONMENT PROTECTION ACT 1985
CONTROL OF PESTICIDES REGULATIONS 1986
(S.I. 1986 No. 1510):
APPROVAL FOR OFF-LABEL USE OF AN APPROVED PESTICIDE PRODUCT

This approval provides for the use of the product named below in respect of crops and situations, other than those included on the product label. Such 'off-label use' as it is known is at all times done at the user's choosing, and the commercial risk is entirely his or hers.

The conditions below are statutory. They must be complied with when the off-label use occurs. Failure to abide by the conditions of approval may constitute a breach of that approval, and a contravention of the Control of Pesticides Regulations 1986. The conditions shown below supersede any on the label *which would otherwise apply*.

Level and scope:	In exercise of the powers conferred by regulation 5 of the Control of Pesticides Regulations 1986 (SI 1986/1510) and of all other powers enabling them in that behalf, the Minister of Agriculture, Fisheries and Food and the Secretary of State, hereby jointly give full approval for the advertisement, sale, supply, storage and use of
Product name:	Dow shield containing
Active ingredient:	200 g/l clopyralid
Marketed by:	DowElanco Ltd under MAFF No. 05578 subject to the conditions relating to off-label use set out below:

Date of issue:	15 July 1992
Date of expiry:	(unlimited subject to the continuing approval of MAFF 05578)
Field of use:	ONLY AS A FORESTRY HERBICIDE
Crops:	Coniferous and broadleaved trees
Maximum individual dose:	1 litre product/hectare
Maximum number of treatments:	Two per year
Operator protection:	(1) Engineering control of operator exposure must be used where reasonably practicable in addition to the following personal protective equipment: Operators must wear suitable protective gloves and face protection (faceshield) when handling the concentrate.
	(2) However, engineering controls may replace personal protective equipment if a COSHH assessment shows they provide an equal or higher standard of protection.
Other specific restrictions:	(1) This product must only be applied if the terms of this approval, the product label and/or leaflet and any additional guidance on off-label approvals have first been read and understood.
	(2) Livestock must be kept out of treated areas for at least

7 days following treatment and until poisonous weeds such as ragwort have died and become unpalatable.

Signed J Micklewright
(Authorised signatory)

Date 15 July 1992

Application Reference Number: COP 91/00752

THIS NOTICE OF APPROVAL IS NUMBER 0757 of 1992

ADVISORY INFORMATION

This approval relates to the use of Dow Shield on coniferous and broadleaved trees. Overall spray may result in crop damage. Tree guards should be used to prevent foliar contamination. When spraying with a knapsack sprayer use 1 part product to 240 parts water. Do not spray to run-off.

4.4 Dalapon/Dichlobenil

Product

Fydulan G	10% w/w dalapon 6.76% w/w dichlobenil	(Zeneca, marketed by Nomix-Chipman)

This product is approved at the time of writing, but manufacture is likely to cease in the future.

Description

A soil acting residual herbicide in granular form. All annual weeds, seedling perennials, most perennial grasses, rushes, bracken and some perennial broadleaved weeds are controlled.

Crop tolerance

SS, NS, (not Christmas trees), DF, CP, LP, SP, oak, beech and sycamore are tolerant. Birch, cherry, holly, willow and rowan that have been established at least 2 years in their final planting position, can also be treated.

All other species may be damaged if the product is applied overall, or to the surrounding soil.

Product rate

35–40 kg/ha, depending on weed density. This can be increased to 45–65 kg/ha, for trees established at least 2 years in their final planting position.

Methods of application

Pre-plant or post-plant, overall, spot or band application
Apply evenly over area to be treated, using the pepperpot applicator.
Refer to Section 11 for details of applicators and correct calibration.

Timing of application

Apply before the end of winter. For pre-planting applications, allow at least 2 months before planting in the treated area. For post planting uses, apply as soon as possible after the soil has settled, then as required in subsequent years.

Additional information

1. **Weed control**

a. Residual activity in the soil should be 3–6 months, although this may be less in warm conditions.

b. All soil types can be treated.

c. Ensure granules do not blow into areas not intended for treatment.

d. The crop safety margin is narrow with this product – great care must be taken to avoid overdosing, particularly around the base of trees. Users should consider alternative herbicides if they have any doubt about their ability to apply this granular product accurately at the specified dose rates.

e. Weed control will be lessened by disturbance of treated soil.

f. Do not apply when crop foliage is wet – granules may lodge on leaves and cause damage.

g. Do not apply on snow covered, frozen or waterlogged sites.

2. Protective clothing

Read the product label for protective clothing and equipment requirements, and check there are no items required in addition to the Forestry Commission recommendations in Section 10.

3. Special precautions

a. Irritating to eyes and skin – avoid contact.

b. Harmful to fish – avoid contamination of ponds, watercourses or ditches with the chemical or used containers.

The label on the herbicide container has been designed for your protection – ALWAYS READ THE INSTRUCTIONS ON THE LABEL.

4.5 2,4-D

Product

Dicotox Extra 400 g/litre 2,4-D (Rhone Poulenc Environmental)

Description

2,4-D is a plant growth regulating herbicide to which many herbaceous and woody broadleaved species are susceptible. Grasses are unaffected at the product rate given. It is absorbed mainly through aerial parts of the plant but also through the roots.

One month should elapse between treatment and subsequent planting.

Crop tolerance

In general, overall sprays should not be made, but directed to avoid contact with foliage, immature bark or leaders.

SS, NS and OMS are moderately tolerant of overall sprays provided leader growth has hardened. Hardening can occur as early as the end of July or may be delayed until the end of October in some locations and seasons.

SP, CP, DF, WRC, GF and NF are rather less tolerant and spray should be directed away from foliage, particularly leading shoots.

LP, WH and larches are sensitive.

Broadleaves are very sensitive to 2,4-D.

Hot weather at the time of spraying may increase crop damage, due to the risk of volatilisation.

Product rate

8 litres/hectare – in young conifer plantations (>1 m in height).
13 litres/hectare – in established trees.

Methods of application

Pre-plant (overall, band or spot) or *post-plant (directed, spot or band)*

Handheld equipment

Knapsack sprayer at MV.
Forestry spot gun at LV.
Refer to Section 11 for details of applicators and correct calibration.

Tractor mounted equipment

Boom sprayer at MV.

Timing of application

Apply when weeds are in active growth, at the early part of the season. All crop species are likely to be sensitive at this time of year. Only directed sprays should be used.

Additional information

1. **Weed control**

a. Broadleaved plants are very sensitive to 2,4-D especially in warm weather, avoid spraying when wind would cause drift and damage to neighbouring crops or trees.

b. To reduce the risk of crop damage trees should be 1 metre tall to minimise contact of spray mixture and leading shoots.

c. Volatilisation can cause serious damage to neighbouring crops – if possible avoid applications in hot weather.

2. **Protective clothing**

Read the product label for protective clothing and equipment requirements and check there are no items required in addition to the Forestry Commission recommendations in Section 10.

3. **Special precautions**

a. Special precautions are required in water catchment areas to avoid water taint – see Section 3.5.

b. If possible, 2,4-D should not be applied in areas visited regularly by the public in significant numbers, and where edible fruits or plants are likely to be exposed to spray. If, however, spraying does take place in such areas appropriate warning notices must be erected. Such signs should remain in position as long as treated fruit appear wholesome.

c. Harmful to fish; do not contaminate ponds, watercourses or ditches with the chemical or used container.

The label on the herbicide container has been designed for your protection – ALWAYS READ THE INSTRUCTIONS ON THE LABEL.

4.6 2,4-D/Dicamba/Triclopyr

Product

Broadshot 200:85:65 g/litre 2,4-D/dicamba/triclopyr (Cyanamid)

Description

A translocated herbicide which controls a wide range of annual and perennial herbaceous weeds, as well as woody weeds. Most grasses are resistant, although some may be suppressed.

Crop tolerance

Conifers may be tolerant when fully dormant but contact from spray with crop foliage or rooting zone should be avoided.

All broadleaves are susceptible to any contact.

Product rate

Herbaceous weeds 3–5 litres/hectare, depending on weed spectrum (refer to label).

For weedwiper application, mix 1 part product to 8 parts water.

Methods of application

Pre-plant (band, spot or overall), post-plant directed spray

Tractor mounted equipment

Boom sprayer at LV or MV (150–400 l/ha).
Use the higher volume rate for larger, denser weed populations.

Handheld equipment

Weedwiper for direct application – uncalibrated, see product rate section.
Knapsack sprayer at LV or MV.
Forestry spot gun at LV.
Refer to Section 11 for details of applicators and correct calibration.

Timing of application

Best control is achieved by applications after weed leaves are fully developed, but before flower buds open, in spring or summer.

Additional information

1. **Weed control**

a. Do not apply when rain is imminent, or in periods of very hot or cold weather.

b. Volatilisation can cause serious damage to neighbouring crops – if possible avoid applications in hot weather.

c. Some woody species may require more than one application – see Sections 7–8.

d. After spraying herbaceous vegetation, 3 months delay should be allowed before replanting.

2. Protective clothing

Read the product label for protective clothing and equipment requirements, and check there are no items required in addition to the Forestry Commission recommendations in Section 10.

3. Special precautions

i. Irritating to eyes and skin, avoid contact. Harmful if swallowed.

ii. The herbicide can be dangerous to fish – do not contaminate ponds, watercourses or ditches with the chemical or used containers.

The label on the herbicide container has been designed for your protection – ALWAYS READ THE INSTRUCTIONS ON THE LABEL.

4.7 Glufosinate ammonium

Products

Challenge	150 g/l glufosinate ammonium (Hoechst/Agr Evo)
Harvest	150 g/l glufosinate ammonium (Hoechst/Agr Evo)

Description

A non-translocated, foliar acting herbicide. Glufosinate ammonium controls a broad spectrum of annual and perennial grasses and broadleaved weeds. Gradual chlorosis over the first 2–3 days is followed by a withering of treated foliage over the next 10–14 days. No translocation to rhizomes or stolons occurs – deep rooted species may require repeat applications for a complete kill. Activity is greatest under warm, moist conditions, when weeds are actively growing. Breakdown of the product is rapid upon contact with the soil.

Crop tolerance

All species will be damaged to some degree if spray is allowed to contact with dormant or green buds, damaged or green bark, and foliage.

Product rate

Annual weed species – 3 litres per hectare.
Perennial weed species – 5 litres per hectare.

Methods of application

Pre-plant (overall, spot or band) or post-plant directed spray

Tractor mounted equipment

Boom sprayer at MV, 200–500 litres/ha – use the higher rate when weeds are large.

Handheld equipment

Knapsack sprayer at MV.
Refer to Section 11 for details of applicators and correct calibration.

Timing

May be applied at any time of year when weeds are actively growing, but best control will be achieved between 1 March and 30 September. Maximum number of applications is two per year.

Additional information

1. **Weed control**

a. Do not mix with any additional wetters.

b. Optimum weed control occurs when weeds have at least two expanded leaves, and are actively growing.

c. Cultivation can take place 4 hours after application.

d. Do not apply to wet foliage, or if rain is imminent.

2. Protective clothing

Read the product label for protective clothing and equipment requirements, and check there are no items required in addition to the Forestry Commission recommendations in Section 10.

3. Special precautions

a. Irritating to eyes – avoid contact.

b. Harmful if swallowed, or in contact with skin.

c. Glufosinate ammonium can be harmful to fish – do not contaminate ponds, watercourses or ditches with the chemical or used containers.

The label on the herbicide container has been designed for your protection – ALWAYS READ THE INSTRUCTIONS ON THE LABEL.

4.8 Glyphosate

Products

Roundup Biactive	360 g/litre glyphosate (Monsanto)
Roundup Pro Biactive	360 g/litre glyphosate (Monsanto)

Description

A translocated herbicide taken up by the foliage and conveyed to the roots. It causes chlorosis and eventual death of leaves and then kills roots and shoots. Symptoms usually become apparent after about 7 days following treatment, but may take longer to show if vegetation growth is slow.

Glyphosate controls a wide range of weeds including grasses, herbaceous broadleaved weeds, bracken, heather and woody weeds. When applied late in the growing season, the main effect is obtained in the following year.

On contact with the soil glyphosate is quickly inactivated. Planting can be carried out 7 days after treatment, and a minimum of 7 days should be allowed before cultivation and the breaking up of rhizomes and roots.

Crop tolerance

SS, NS, SP, CP, LP, WRC and LC will tolerate overall sprays provided leader growth has hardened. Hardening can occur as early as the end of July or may be delayed until the end of October in some locations and seasons. Hardening may be indicated when the leader changes from bright green to a straw colour. This may occur after buds have hardened. To avoid damage to lammas growth, herbicide sprays must be directed away from leaders. During the active growing season trees must be guarded or the spray directed to avoid contact with the crop.

DF and NF as above but much more sensitive – do not spray in spring.

Broadleaved trees, larch and other conifers will not tolerate overall applications; always use a guard, weedwiper or a directed spray to avoid contact with the foliage and immature bark of crop trees.

Product rate

1. *Pre-plant (overall, band or spot)*
   4.0–5.0 litres/hectare

2. *Post-plant (directed)*
   4.0 litres/hectare
   For direct applications using the weedwiper, use 1 part product diluted in 2 parts water. Red dye (available from Hortichem Ltd.) can be used to mark areas if required.
3. *Post-plant (overall, dormant season)*
   Lowland areas 1.5 litres/hectare
   Upland areas 2.0 litres/hectare

Methods of application

1. *Pre-plant (overall, band or spot application)*

Tractor mounted equipment
Boom sprayer at LV or MV.
Ulvaforest low speed rotary atomiser at VLV.

Handheld equipment
Knapsack sprayer at MV.
Knapsack sprayer at LV.
Herbi low speed rotary atomiser at VLV.

2. *Post-plant (directed, overall, band or spot application)*

Tractor mounted equipment
Boom sprayer at LV or MV.
Ulvaforest low speed rotary atomiser at VLV.

Handheld equipment
Knapsack sprayer at MV.
Knapsack sprayer at LV.
Herbi low speed rotary atomiser at VLV.
Forestry spot gun for spot application at LV.
Weedwiper for direct application, uncalibrated (see Product rate).

Refer to Section 11 for details of applicators and correct calibration.

Timing of application

Glyphosate can be applied at any time of year when vegetation is actively growing but is most effective on broadleaved weeds when they are close to flowering but before senescence.

When crop trees are present this will influence the timing and/or method of application (see Crop tolerance).

Additional information

1. **Weed control**

a. Glyphosate applied later than June will be too late to lessen the effect of weed competition in the current season.

b. Glyphosate is most effective when relative humidity is high and the air is warm (e.g. 15°C+).

c. Reduced weed control may result when weeds are under stress, for example from frost or drought.

d. Direct application using the weedwiper, achieves maximum control when the vegetation is actively growing and under 0.3 m in height. In taller vegetation and where a large number of seed heads are present the degree of control will be reduced. Care must be taken to avoid low tree branches amongst weeds.

e. Heavy rainfall within 24 hours of application may reduce the herbicide's effectiveness by preventing sufficient foliar absorption. The addition of Mixture B at 2% of final spray volume will improve reliability in these circumstances. The addition of Mixture B will reduce crop tolerance and it must only be used pre-plant or post-plant as a directed spray. Do not use Mixture B with rotary atomisers.

f. Diluted product may denature after 2 to 3 days if clean water is not used. Where possible use tap water as the diluent and only mix sufficient for the day's programme.

g. Do not overall spray Christmas trees or trees grown for ornamental purposes.

h. Late June applications are likely to be the most effective treatment for *Calamagrostis epigejos, Holcus mollis* and *Molinia caerulea.* September applications will probably give the best control of *Deschampsia flexuosa.*

2. **Protective clothing**

Read the product label for protective clothing and equipment requirements and check there are no items required in addition to the Forestry Commission recommendations in Section 10.

3. **Special precautions**

a. Unlike the additional glyphosate products listed at the end of this Section, Roundup Pro Biactive and Roundup Biactive DO NOT require a warning on the product label about harmful effects to aquatic life, or about irritant effects to operators. However, the addition of Mixture B makes any application potentially harmful to aquatic life, and a potential irritant to operators.

b. Do not contaminate ponds, watercourses or ditches with the chemical or used containers.

   The label on the herbicide container has been designed for your protection – ALWAYS READ THE INSTRUCTIONS ON THE LABEL.

Additional glyphosate products with full forestry approval

Product	Formulation
Barclay Gallup	360 g/litre glyphosate (Barclay)
Barclay Gallup Amenity	360 g/litre glyphosate (Barclay)
Clayton Glyphosate	360 g/litre glyphosate (Clayton)
Clayton Swath	360 g/litre glyphosate (Clayton)
Glyphogan	360 g/litre glyphosate (PBI)
Glyphosate–360	360 g/litre glyphosate (Top Farm)
Helosate	360 g/litre glyphosate (Helm)
Hilite	144 g/litre glyphosate (Nomix-Chipman) – CDA formulation
Outlaw	360 g/litre glyphosate (Barclay)
Portman Glyphosate 360	360 g/litre glyphosate (Portman)
Roundup	360 g/litre glyphosate (Monsanto)
Roundup	360 g/litre glyphosate (Schering/Agro Evo)
Roundup Biactive Dry	42.6% w/w glyphosate (Monsanto)
Stacato	360 g/litre glyphosate (Unicrop)
Stefes Glyphosate	360 g/litre glyphosate (Stefes)
Stefes Kickdown 2	350 g/litre glyphosate (Stefes)
Stetson	360 g/litre glyphosate (Monsanto)
Stirrup	144 g/litre glyphosate (Nomix-Chipman) – CDA formulation

These products may have different conditions of use regarding operator and environmental safety – refer to the product label for further guidance.

4.9 Imazapyr

Product

Arsenal 50F 50 g/litre imazapyr (Cyanamid, marketed by Nomix-Chipman)

Description

A translocated and residual herbicide which will control a wide range of established annual and perennial grass and herbaceous broadleaved weeds, and will provide long term residual control of germinating seedlings. Control of woody weeds is likely to be good, although there are no manufacturer's label recommendations at present. Imazapyr is absorbed through roots and foliage, then rapidly translocated throughout the plant. Treated plants stop growing soon after application, but chlorosis and tissue necrosis may not be apparent until up to 2 weeks after application – complete kill may take several weeks.

Crop tolerance

Imazapyr should only be used as a pre plant treatment. The only tolerant species according to the product label are SS, LP and CP. Limited FC trials suggest JL, DF and NS may also be tolerant. The trees must be at least 2 year old nursery stock, and 5 months must elapse between treatment and subsequent planting. All other species may be damaged.

Product rate

7.5 litres/hectare Annual and perennial seedlings and small plants.
10 litres/hectare Established annuals and grasses.
15 litres/hectare Dense and deep rooted perennial grasses.

Rates vary depending on weed species present – see the product label. Generally, the higher rate is required for well established perennial species.

Methods of application

Pre-plant (overall or band or spot)

Tractor mounted equipment

Boom sprayer at LV–MV (200 l/ha).
Ulvaforest low speed rotary atomiser at VLV.

Handheld equipment

Knapsack sprayer at LV–MV (200 l/ha).
Forestry spot gun at LV.
Refer to Section 11 for details of applicators and correct calibration.

Timing of application

Apply from July–October, when weeds are actively growing.

Additional information

1. **Weed control**

a. Residual weed control will be reduced if cultivation takes place after spraying. Ideally, a site should be sprayed after cultivation has occurred, and when any weeds present have started to re-grow. When this is not possible, a period of at least a week should elapse after treatment, prior to cultivation.

b. Do not use imazapyr on soils that may later be used for growing desirable plants, except for the coniferous species SS, CP and LP.

c. Do not use on conifer nursery beds.

d. Do not use near desirable trees or shrubs, nor in areas into which their roots may extend or in locations where the chemical may be washed or moved into contact with their roots.

e. Do not apply in windy conditions, or using high pressures producing a fine spray liable to drift.

f. The length of residual weed control given has not been fully evaluated, but indications are that in some circumstances weed suppression may still occur up to two seasons after planting. However, this may vary from site to site, and there are no manufacturer's recommendations on this subject at present.

g. Results from trials suggest that planted trees are unlikely to be damaged from root contact with treated vegetation, or from leachates from dead vegetation that is subsequently cultivated. However, imazapyr may be very mobile in soil.

2. **Protective clothing**

Read the product label for protective clothing and equipment requirements, and check there are no items required in addition to the Forestry Commission recommendations in Section 10.

3. **Special precautions**

a. Irritating to eyes – avoid contact.

b. Do not contaminate ponds, watercourses or ditches with the chemical or used containers.

The label on the herbicide container has been designed for your protection – ALWAYS READ THE INSTRUCTIONS ON THE LABEL.

4.10 Isoxaben

Product

Gallery 125 125 g/litre isoxaben (DowElanco)
Flexidor 125 125 g/litre isoxaben (DowElanco)

Description

A soil acting residual herbicide giving pre-emergent herbaceous broadleaved weed control. Isoxaben is most effective on moist soil, free of clods, brash and weeds. It is probably most appropriate therefore for new planting sites that have been fully cultivated, or in nursery situations.

Crop tolerance

DF, EL, JL, NF, NS, SP, alder, ash, beech, birch, cherry, oak, poplar, sweet chestnut, sycamore and willow are tolerant, providing they have good root development.

Product rate

2.0 litres/hectare (or 0.5 litres/hectare for the old 500 g/litre formulation).

Methods of application

Pre- or post-plant (overall, band or spot)

Tractor mounted equipment

Boom sprayer at MV.

Handheld applicators

Knapsack sprayers at MV.
Forestry spot gun at LV.

Refer to Section 11 for details of applicators and correct calibration.

Timing of application

Best results will be obtained from applications to newly planted sites, prior to weed emergence. Two applications may be made per year.

Additional Information

1. **Weed control**

a. Isoxaben will have little effect on established weeds.

b. Any disturbance of the soil in a treated area, such as a planting operation, will reduce weed control.

c. Isoxaben can be applied safely to sandy, light, medium and heavy soils. Activity will be reduced on soils with more than 10% organic matter content.

d. For new plantings in forestry situations, a tank mix with propyzamide will allow pre-emergence control of grasses as well as herbaceous weeds. For Farm Forestry situations, refer to Section 9.

2. **Protective clothing**

   Read the product label for protective clothing and equipment requirements, and check there are no items required in addition to the Forestry Commission's recommendations in Section 11.

3. **Special precautions**

   Do not contaminate surface waters or ditches with the chemical or used container. The label on the herbicide container has been designed for your protection – ALWAYS READ THE INSTRUCTIONS ON THE LABEL.

4.11 Propyzamide

Product

Kerb Granules	4% w/w propyzamide (PBI/Rohm + Haas)
Kerb 50W	50 % w/w propyzamide: wettable powder (PBI/Rohm + Haas)
Kerb Flowable	400 g/litre propyzamide (PBI/Rohm + Haas)

Description

A soil acting herbicide which slowly volatilises in cold soil and is taken up by germinating weeds and through the roots of existing weeds, especially grasses. Most grasses, and some herbaceous broadleaved weeds, are susceptible from germination to the true leaf stage, but herbaceous broadleaved weeds which emerge late in the season will only be partially controlled.

Propyzamide slowly breaks down in the soil, lasting for 3–6 months.

Crop tolerance

All commonly planted forest tree species are tolerant.

Kerb Granules

Product rate

Apply at 38.0 kg of granules per treated hectare.

Methods of application

Pre- or post-plant (overall, band or spot application)

Apply through the Pepperpot or Tree Mate kerb granule applicator.

Kerb 50W

Product rate

Apply as 3 kg of product per treated hectare in water.

Methods of application

Pre- or post-plant (overall, band or spot application)

Tractor mounted equipment

Boom sprayer at MV.

Handheld equipment

Knapsack sprayer at MV.
Knapsack sprayer with 'VLV' nozzle at LV.
Forestry spot gun for spot application at LV.
For best results mix the powder into a smooth paste with a small

amount of water, slowly add more water, stirring all the time, until the mix is sloppy, then add the remaining water up to full dilution.

Kerb Flowable

Product rate

Apply at 3.75 litres of product per treated hectare.

Methods of application

Pre- or post-plant (overall, band or spot application)

Tractor mounted equipment

Boom sprayer at MV.

Ulvaforest low speed rotary atomiser at VLV.

Handheld equipment

Knapsack sprayer at MV.

Knapsack sprayer with VLV nozzle at LV.

Herbi low speed rotary atomiser at VLV.

Forestry spot gun at LV.

When applying Kerb Flowable via the Herbi or ULVA forest a non-ionic wetting agent (such as Agral) should be added at the rate of 0.5% of final spray volume.

Refer to Section 11 for details of applicator and correct calibration.

Timing of application

Kerb 50W and Kerb Flowable:

Apply from 1 October to 31 January, north of a line from Aberystwyth to London.

Apply from 1 October to 31 December south of a line from Aberystwyth to London, and on peat or peaty gley soils.

Kerb Granules:

Apply from 1 October to end of February, north of a line from Aberystwyth to London.

Apply from 1 October to end of January, south of a line from Aberystwyth to London, and on peat or peaty gley soils – see Additional Information 1.b.

Additional information

1. **Weed control**

a. Although propyzamide can be used very effectively for pre-planting applications, the effect of the herbicide applied in winter cannot often be seen by the time of normal planting.

b. Organic soils and surface debris decrease the activity of propyzamide – treatment of soils with a depth of peat greater than 10 cm is not recommended.

c. The following grasses show some resistance to propyzamide.

Dactylis glomerata	(cocksfoot)
Holcus mollis	(creeping soft grass)
Calamagrostis epigejos	(small reed grass)

d. Planting can be carried out immediately pre- or post-treatment.

e. Propyzamide is very insoluble and requires rainfall to move it into the soil where it can be taken up by the roots of weeds. It works best on a cold, firm, moist tilth.

f. Warm, dry conditions after application may reduce the level of weed control.

2. **Protective clothing**

Read the product label for protective clothing and equipment requirements and check there are no items required in addition to the Forestry Commission recommendations in Section 10.

3. **Special precautions**

Propyzamide concentrate can be harmful to fish; do not contaminate ponds, watercourses or ditches with the chemical or used containers.

The label on the herbicide container has been designed for your protection – ALWAYS READ THE INSTRUCTIONS ON THE LABEL.

4.12 Triclopyr

Products

Timbrel	480 g/litre triclopyr (DowElanco)
Garlon 4	480 g/litre triclopyr (DowElanco)
Chipman Garlon 4	480 g/litre triclopyr (Nomix-Chipman)

Description

A plant growth regulating herbicide which is rapidly absorbed, mainly through the foliage, but also by roots and stems. Once inside the plant it is readily translocated. Triclopyr is effective against most herbaceous and woody weeds.

Grasses sometimes show some yellowing following spraying operations but this is quickly outgrown.

In the soil triclopyr is broken down fairly rapidly by microbial action. Planting should be deferred for at least 6 weeks after application.

Crop tolerance

SS and NS will tolerate overall sprays up to 4 l/ha provided leader growth has hardened. Hardening can occur as early as the end of July or may be delayed until October in some locations and seasons. To avoid damage to lammas growth, herbicide sprays must be directed away from leaders. During the active growing season trees must be guarded or the spray directed to avoid contact with the crop.

DF and pines are rather more sensitive than spruces with occasional leader damage if sprayed overall.

Larch, other conifers and broadleaves are severely damaged by overall sprays at any time of year. Late September application will be tolerated if applied with great care to avoid the foliage of crop trees.

Product rate

Herbaceous weeds 2.0 litres per hectare.

Methods of application

1. *Pre-plant (overall application)*

Tractor mounted equipment

Tractor mounted boom sprayer at MV.

Handheld equipment

Knapsack sprayer at MV.

2. *Post-plant (directed, overall, band or spot application)*

Handheld equipment

Knapsack sprayer at MV.
Forestry spot gun at LV.

Refer to Section 11 for details of applicators and correct calibration.

4

Timing of application

The timing of overall applications will be determined by the growth stage of the crop trees (see Crop tolerance), but to be effective the operation must take place after crop trees have hardened but before the target species show signs of senescence.

Directed applications can take place at any time between June–September provided the target species shows no signs of senescence and adequate precautions are taken to protect the crop species.

Additional information

1. **Weed control**

a. Volatilisation can cause serious damage to neighbouring crops – if possible avoid applications in hot weather.

b. Rainfall within 2 hours of application may reduce the herbicide's effectiveness by preventing sufficient foliar absorption.

c. A minimum interval of 6 weeks is required between application of triclopyr and planting.

2. **Protective clothing**

Read the product label for protective clothing and equipment requirements and check there are no items required in addition to the Forestry Commission recommendations in Section 10.

3. **Special precautions**

a. Triclopyr is harmful in contact with skin and irritating to eyes – avoid contact.

b. Triclopyr should not be applied via controlled droplet applicators.

c. Triclopyr is dangerous to fish – do not contaminate ponds, watercourses or ditches with the concentrate, spray drift or used containers.

d. Do not apply with tractor mounted boom sprayers within 250 m of ponds, ditches or watercourses.

   The label on the herbicide container has been designed for your protection – ALWAYS READ THE INSTRUCTIONS ON THE LABEL.

5 Bracken

5.1 General

Bracken competes strongly with young trees for light during the latter part of the growing season. At the end of the year it collapses and can smother and flatten small trees with its weight, particularly if snow lies on top of them both. Bracken is rarely sufficiently advanced in spring to afford protection from frosts and so is probably not worth retaining for this purpose.

Ploughing does give some control of bracken for the first season, but on sites where bracken is vigorous, the stems on either side of the plough ridge will overgrow conifers.

If a crop is present, chemical control must be followed at least one month later by hand cutting before the fronds collapse on to the trees and cause damage. Whenever possible, herbicide should be applied pre-planting to avoid this problem.

The herbicides that can be used for bracken control are:

asulam, dicamba, imazapyr, glyphosate.

Asulam, glyphosate and imazapyr are translocated from the fronds to the rhizomes where they have a herbicidal effect. The result is that frond growth the following year is prevented or retarded.

Asulam gives good control of bracken but no other weeds are controlled. Conifer tolerance is high.

Dicamba is applied to the soil or bracken litter, then washed down to the rhizomes by rainfall. Control is best in dense bracken infestation.

Glyphosate gives adequate control, but can cause damage to crop species, particularly if used in mid-summer.

Imazapyr will control bracken, but can only be used as a pre-plant treatment.

On sites with a mixture of bracken and other weed types, glyphosate or imazapyr may be the best choice due to their broad weed control spectrum. Dicamba will give good pre-plant control where bracken is the only weed present. Asulam may be the most appropriate post-plant treatment where bracken is the primary weed problem.

5.2 Asulam

Product

Asulox 400 g/litre asulam (Rhone-Poulenc Agriculture)

Description

Asulam is a post-emergence, translocated herbicide which is taken up by the foliage and translocated to the rhizomes. Herbicidal symptoms are virtually absent in the year of spraying but growth of the bracken is then retarded or fails entirely the following season. Control may last from 1 to 4 years, or more depending on the rate applied and the date of application.

Some grasses and herbaceous vegetation (particularly docks) may be damaged at the higher product rate.

Crop tolerance

Mature species of SS, NS, DF, GF, JL, CP, SP, beech, birch, elm, poplar, holly and rowan are tolerant of applications. However, young trees may exhibit chlorosis and slight check in growth if directly sprayed whilst actively growing.

WH and willow are more susceptible – all spraying of these species should be avoided.

At least 6 weeks should be left between treatment and subsequent planting.

Product rate

Apply 5 litres (early applications) – 10 litres (late season applications) of product per hectare diluted in water.

The higher rate will give increased duration of bracken control.

Methods of application

Pre-plant (overall, or band application) or post-plant (directed, overall or band)

Tractor mounted equipment

Boom sprayer at MV.
Ulvaforest low speed rotary atomiser at VLV.

Handheld equipment

Knapsack sprayer at MV.
Knapsack sprayer at LV.
ULVA high speed rotary atomiser at VLV.

Forestry spot gun at LV.

Aerial applications (check requirements of Control of Pesticides Regulations) – apply Asulox in 55 litres of water per hectare using either Delavan 'Raindrop' or conventional nozzles.

Refer to Section 11 for details of applicators and correct calibration.

Timing of application

a. General

Apply early June to late August (early August in northern Britain). Best results are obtained by application just as the frond tips have unfurled and formed an almost complete canopy. Treatment at this stage may reduce the need to hand-cut the dead stems at the end of the growing season.

b. Pre-planting

Areas to be planted should be treated the summer before planting. Spraying can take place in standing timber, prior to felling, but damage to treated bracken from harvesting operations may lead to some reduction in weed control.

c. Post-planting

Treatement can either be made to the undisturbed canopy, or bracken can be cut early in the season, and the late-season re-growth sprayed. This latter approach means applications can take place when tree growth is more likely to have ceased.

Additional information

1. Weed control

a. To obtain optimum control of bracken a minimum of 14 days should elapse after treatment before cutting the bracken down or ploughing the ground.

b. Access for spraying should be made by pushing fronds aside and not by cutting.

c. Heavy rainfall within 24 hours of application may reduce the herbicide's effectiveness by preventing sufficient foliar absorption.

d. Reduced weed control may result when weeds are under stress, for example from frost or drought, or when application is made during conditions of high temperature and low humidity.

e. For best results spray before the onset of senescence.

2. **Protective clothing**
   Read the product label for protective clothing and equipment requirements and check there are no items required in addition to the Forestry Commission recommendations in Section 10.
3. **Special precautions**
   Asulam concentrate can be harmful to fish; do not contaminate ponds, watercourses or ditches with the chemical or used containers.

   The label on the herbicide container has been designed for your protection – ALWAYS READ THE INSTRUCTIONS ON THE LABEL.

5.3 Dicamba

Product

Tracker 480 g/litre dicamba (PBI)

Description

Dicamba is a translocated soil acting herbicide for bracken control. The dilute spray is applied to the soil or bracken litter in a narrow band which must be washed into the soil by rainfall. It is then absorbed by rhizomes and translocated within the plant. Some herbaceous broadleaved weeds and grasses may also be suppressed or killed.

Crop tolerance

All species will be killed or severely damaged if applications take place within the rooting zone of the crop, or if the chemical comes into contact with foliage. In addition, applications within the rooting zone of mature trees must be avoided.

Product rate and method of application

5 litres of product will treat 1 hectare at 2 m crop spacing.

Pre- or post-plant (band)

Handheld applicators

Use on row widths between 1.8 and 2.2 m.

Application rate at walking speed of one metre per second using LURMARK GREY METERING DISC (without the swirl plate)

Calibrated output in ml per minute	ml of TRACKER to be added to each litre of clean water		Metres treated per litre of spray mixture
	Moderate frond suppression	Frond control	
300	190	250	195
360	150	200	158
420	130	170	133
480	110	140	114
540	95	125	100
600	85	110	90
660	75	100	82
720	70	90	75
780	65	85	69
840	60	80	65
900	55	75	60

Apply using a knapsack sprayer fitted with a Lurmark Grey metering disc, without the swirl plate, at 1.25–2.5 bar pressure, with the lance held 10–20 cm above the ground, to achieve a 2–5 cm BAND MIDWAY BETWEEN CROP ROWS.

Determine output from the sprayer in ml/minute, then refer to the preceding table to obtain the correct dilution rate.
To determine dilution rates when walking speed varies from 1 m/second, use the following formula:
Corrected ml of product/litre water = ml/litre product from table × measured walking speed [metres/second].

Timing of application

Apply from March to early May, before frond emergence. Apply during the season of planting, or up to 2 years after planting. Care must be taken to avoid applications to rooting zone of crop trees in post-plant treatments.

Additional information

1. **Weed control**

a. Dicamba is suitable for use in plantings where the row width is between 1.8 and 2.2 m. When used as recommended, the roots of crop trees should not enter the zone of chemical activity.

b. Where trees are planted on a ploughed furrow, applications should be delayed until the year after planting, to ensure the bracken rhizomes interrupt the band treatment.

c. Dicamba is most effective in dense bracken infestations.

d. Do not use where there is a risk of run-off, for example on light or shallow soils and steep slopes.

2. **Protective clothing**

Read the product label for protective clothing and equipment requirements and check there are no items required in addition to the Forestry Commission recommendations in Section 10.

3. **Special precautions**

Do not contaminate surface water, streams or ditches with the concentrate or used containers.

The label on the herbicide container has been designed for your protection – ALWAYS READ THE INSTRUCTIONS ON THE LABEL.

5.4 Glyphosate

Products

Roundup Biactive	360 g/litre glyphosate (Monsanto)
Roundup Pro Biactive	360 g/litre glyphosate (Monsanto)

Description

A translocated herbicide taken up by the foliage and conveyed to the rhizomes. It causes chlorosis and eventual death of fronds and prevents regrowth.

Glyphosate controls a wide range of weeds including grasses, herbaceous broadleaved weeds, bracken, heather and woody weeds. It is particularly effective for bracken in mixture with these other weed types. On bracken some dieback of foliage can be expected in the year of application. In the following season rhizomes fail to send out fronds.

On contact with the soil glyphosate is quickly inactivated. Planting can be carried out 7 days after treatment, and a minimum of 7 days should be allowed before cultivation and the breaking up of rhizomes and roots.

Crop tolerance

SS, NS, SP, CP, LP, WRC and LC will tolerate overall sprays provided leader growth has hardened. Hardening can occur as early as the end of July or may be delayed until the end of October in some locations and seasons. Hardening may be indicated when the leader changes from bright green to a straw colour. This may occur after buds have hardened. To avoid damage to lammas growth, herbicide sprays must be directed away from leaders. During the active growing season trees must be guarded or the spray directed to avoid contact with the crop.

DF and NF as above but much more sensitive.

Broadleaved trees, larch and other conifers will not tolerate overall applications; always use a guard, or a directed spray to avoid contact with the foliage and immature bark of crop trees.

Product rate

Pre-plant (overall or band) or post-plant (directed)	– 3.0 1/ha.
Post-plant (overall, dormant season)	– 2.0 1/ha.

Methods of application

Pre-plant (overall or band application) or post-plant (overall, band or directed application)

Tractor mounted equipment

Boom sprayer at LV or MV or LV.

Ulvaforest low speed rotary atomiser at VLV.

Handheld equipment

Knapsack sprayer at MV or LV.

ULVA high speed rotary atomiser at VLV. For full effect, dilute the herbicide with at least 5 times the volume of water.

Forestry spot gun at LV.

Refer to Section 11 of details of applicator and correct calibration.

Timing of application

Best results will be obtained from applications made during July and August; after frond tips have uncurled but before senescence.

Additional information

1. **Weed control**

a. Glyphosate applied to control bracken in July and August will be too late to lessen the effect of weed competition or smothering in the current season.

b. Glyphosate is most effective when relative humidity is high and the air is warm (e.g. 15°C+).

c. Reduced weed control may result when weeds are under stress from factors such as drought.

d. Heavy rainfall within 24 hours of application may reduce the herbicide's effectiveness by preventing sufficient foliar absorption. The addition of Mixture B at 2% of final spray volume will improve reliability in these circumstances. The addition of Mixture B will reduce crop tolerance and it must only be used pre-plant or post-plant as a directed spray. Do not use Mixture B with rotary atomisers.

e. Diluted product may denature after 2 to 3 days if clean water is not used. Where possible use tap water as the diluent and only mix sufficient for the day's programme.

f. Access for spraying should be made by pushing the fronds aside and not by cutting.

g. A minimum of the one month should elapse after treatment before cutting the bracken down or ploughing the ground.

h. Do not overall spray Christmas trees or trees grown for ornamental purposes.

2. **Protective clothing**

Read the product label for protective clothing and equipment requirements and check there are no items required in addition to the Forestry Commission recommendations in Section 10.

3. **Special precautions**

a. Unlike the additional glyphosate products listed at the end of this Section, Roundup Pro Biactive and Roundup Biactive DO NOT require a warning on the product label about harmful effects to aquatic life, or about irritant effects to operators. However, the addition of Mixture B makes any application potentially harmful to aquatic life, and a potential irritant to operators.

b. Do not contaminate ponds, watercourses or ditches with the chemical or used containers.

The label on the herbicide container has been designed for your protection – ALWAYS READ THE INSTRUCTIONS ON THE LABEL.

5

Additional glyphosate products with full forestry approval

Product	Formulation
Barclay Gallup	360 g/litre glyphosate (Barclay)
Barclay Gallup Amenity	360 g/litre glyphosate (Barclay)
Clayton Glyphosate	360 g/litre glyphosate (Clayton)
Clayton Swath	360 g/litre glyphosate (Clayton)
Glyphogan	360 g/litre glyphosate (PBI)
Glyphosate-360	360 g/litre glyphosate (Top Farm)
Helosate	360 g/litre glyphosate (Helm)
Hilite	144 g/litre glyphosate (Nomix-Chipman) – CDA formulation
Outlaw	360 g/litre glyphosate (Barclay)
Portman Glyphosate 360	360 g/litre glyphosate (Portman)
Roundup	360 g/litre glyphosate (Monsanto)
Roundup	360 g/litre glyphosate (Schering/Agro Evo)
Roundup Biactive Dry	42.6% w/w glyphosate (Monsanto)
Stacato	360 g/litre glyphosate (Unicrop)
Stefes Glyphosate	360 g/litre glyphosate (Stefes)
Stefes Kickdown 2	360 g/litre glyphosate (Stefes)

Stetson	360 g/litre glyphosate (Monsanto)
Stirrup	144 g/litre glyphosate (Nomix-Chipman) – CDA formulation

These products may have different conditions of use regarding operator and environmental safety – refer to the product label for further guidance.

5.5 Imazapyr

Product

Arsenal 50F 50 g/litre imazapyr (Cyanamid, marketed by Nomix-Chipman)

Description

A translocated and residual herbicide which will control a wide range of established weeds including bracken and will provide long term residual control of germinating seedlings. Control of woody weeds is likely to be good, although there are no manufacturer's recommendations at present. Imazapyr is absorbed through roots and foliage, then rapidly translocated throughout the plant. Treated plants stop growing soon after application, but chlorosis and tissue necrosis may not be apparent until up to 2 weeks after application – complete kill may take several weeks.

Crop tolerance

Imazapyr should only be used as a pre-plant treatment. The only tolerant species, according to the product label, are SS, LP and CP. Limited FC trials suggest JL, DF and NS may also be tolerant. The trees must be at least 2 year old nursery stock, and 5 months must elapse between treatment and subsequent planting. All other species may be damaged.

Product rate

7.5 litres/hectare.

Methods of application

Pre-plant (overall or band)

Tractor mounted equipment

Boom sprayer at LV-MV (200 l/ha).
Ulvaforest low speed rotary atomiser at VLV.

Handheld equipment

Knapsack sprayer at LV-MV (200 l/ha).
Forestry spot gun at LV.

Refer to Section 11 for details of applicators and correct calibration.

Timing of application

Best results will be achieved from applications in July or August after full frond emergence, prior to weed senescence.

Additional information

1. **Weed control**

a. Reduced weed control will be reduced if cultivation takes place after spraying. Ideally, a site should be sprayed after cultivation has occurred, and when any weeds present have started to regrow. When this is not possible, a period of at least a week should elapse after treatment, prior to cultivation.

b. Do not use imazapyr on soils that may later be used for growing desirable plants, except for the coniferous species, SS, CP, and LP.

c. Do not use on conifer nursery beds.

d. Do not use near desirable trees or shrubs, nor in areas into which their roots may extend or in locations where the chemical may be washed or moved into contact with their roots.

e. Do not apply in windy conditions, or using high pressures producing a fine spray liable to drift.

f. The length of residual weed control given has not been fully evaluated, but indications are that in some circumstances weed suppression may still occur up to two seasons after planting. However, this may vary from site to site, and there are no manufacturer's recommendations at present.

g. Results from trials suggest that planted trees are unlikely to be damaged from root contact with treated vegetation, or from leachates from dead vegetation that is subsequently cultivated. However, imazapyr may be very mobile in soil.

2. **Protective clothing**
Read the product label for protective clothing and equipment requirements, and check there are no items required in addition to the Forestry Commission recommendations in Section 10.

3. **Special precautions**

a. Irritating to eyes – avoid contact.

b. Do not contaminate ponds, watercourses or ditches with the chemical or used containers.

The label on the herbicide container has been designed for your protection – ALWAYS READ THE INSTRUCTIONS ON THE LABEL.

6 Heather

6.1 General

On sites where the availability of mineral nitrogen limits tree growth, and where the dominant vegetation is heather (*Calluna vulgaris*), nitrogen deficiency may develop in certain species of conifer. This may need to be alleviated by complete spraying to kill the heather.

Sitka spruce is by far the most important species in this context, although NS, WH, GF, NF, DF and CP may also be severely checked by heather competition.

The need to control heather can often be avoided by:

- planting a non-susceptible species;
- burning the heather before ploughing;
- restocking felled areas before the heather has time to invade;
- planting spruce in mixture with SP, larches or LP;
- fertilizing a site.

Herbicides available for heather control are:

- 2,4-D
- glyphosate
- imazapyr

2,4-D may often be the most appropriate herbicide as it can be used selectively during the growing season, although at the rates required to kill heather some damage can be expected on crop species.

Glyphosate can be used in late season when trees are dormant and this may provide the best means of controlling heather amongst trees less than 1 metre in height. It does not taint water and has a very low mammalian toxicity and may therefore be the best choice for water catchment areas. Glyphosate is however likely to cause some damage to crop trees at the rates required to give a good heather kill.

Imazapyr should only be used as a pre-plant treatment. Applications will check heather and give longer term control, but its effects have not yet been fully evaluated.

The level of heather control obtained can vary depending on the vigour of the heather involved. On sites with a high nitrogen status, and rapid turnover of nutrients, heather is usually easier to control.

6.2 2,4-D

Product

Dicotox Extra 400 g/litre 2,4-D (Rhone-Poulenc Environmental)

Description

2,4-D is a plant growth regulating herbicide to which many herbaceous and woody broadleaved species are susceptible. It is absorbed mainly through aerial parts of the plant but also through the roots. The hormonal activity browns the heather shortly after treatment but exhibits its full effect in the following season.

There is a risk to bees through ingestion when spraying heather in flower. This risk can be minimised by good liaison with bee-keepers.

One month should elapse between treatment and subsequent planting.

Crop tolerance

In general, overall sprays should not be made, but directed to avoid contact with foliage, immature bark or leaders.

However, SS, NS and OMS are moderately tolerant of overall sprays provided leader growth has hardened. Hardening can occur as early as the end of July or may be delayed until the end of October in some locations and seasons.

CP, SP, DF, WRC, GF and NF are rather less tolerant and spray should be directed away from foliage, particularly leading shoots.

LP, WH and larches are sensitive.

Broadleaves are very sensitive to 2,4-D.

Product rate

In young conifer plantations (>1 m in height) – 8 l/ha.
In established trees – 13 l/ha.

Methods of application

Pre-plant (overall or band) or post-plant (directed application)

Tractor mounted equipment

Boom sprayer at MV.

Handheld equipment

Knapsack sprayer at MV.
Forestry spot gun at LV.

Refer to Section 11 for details of applicators and correct calibration.

Timing of application

Apply May to August, after crop has hardened off.
Applications earlier than mid-July should only be applied as a directed spray so as to avoid serious crop damage.

Additional information

1. **Weed control**

a. For optimum control of heather apply 2,4-D when the heather is in flower, but because of the risk to bees following application bee-keepers in the locality must be given adequate notice.

b. It is important to get good coverage of heather foliage and stems for satisfactory control.

c. Broadleaved plants are very sensitive to 2,4-D especially in warm weather; avoid spraying when wind would cause drift and damage to neighbouring crops or trees.

d. To reduce the risk of crop damage, trees should be at least 1 metre tall to minimise contact of spray mixture and leading shoots.

e. Volatilisation can cause serious damage to neighbouring crops – if possible avoid applications in hot weather.

f. Heather should be dry at the time of application; heavy rain soon after application may reduce the level of control achieved.

2. **Protective clothing**

Read the product label for protective clothing and equipment requirements and check there are no items require in addition to the Forestry Commission recommendations in Section 10.

3. **Special precautions**

a. Special precautions are required in water catchment areas to avoid water taint – see Section 3.6.

b. If possible 2,4-D should not be applied in areas visited regularly by the public in significant numbers, and where edible fruits or plants are likely to be exposed to spray. If, however, spraying does take place in such areas appropriate warning notices must be erected. Such signs should remain in position as long as treated fruit appear wholesome.

c. Heather flowering will probably be affected by application of 2,4-D earlier than August. Beekeepers should be advised not to site hives on or near the areas to be sprayed.

d. Harmful to fish – do not contaminate ponds, watercourses or ditches with the chemical or used container.

The label on the herbicide container has been designed for your protection – ALWAYS READ THE INSTRUCTIONS ON THE LABEL.

6.3 Glyphosate

Product

Roundup Biactive	360 g/litre glyphosate (Monsanto)
Roundup Pro Biactive	360 g/litre glyphosate (Monsanto)

Description

A translocated herbicide taken up by the foliage and conveyed to the roots. It causes chlorosis and eventual death of leaves and then kills roots and shoots.

Glyphosate controls a wide range of weeds including grasses, herbaceous broadleaved weeds, bracken, heather and woody weeds. The herbicidal effects take some time to develop on heather and the full response is not evident until the following growing season.

On contact with the soil glyphosate is quickly inactivated. Planting can be carried out 7 days after treatment, and a minimum of 7 days should be allowed before cultivation and the breaking up or rhizomes and roots.

Crop tolerance

At the rates used for heather control, spruces and pines are only moderately tolerant to overall sprays of glyphosate, after new growth has hardened. For this reason only directed sprays are recommended post-planting.

Hardening can occur as early as the end of July or may be delayed until the end of October in some locations and seasons. Hardening may be indicated when the leader changes from bright green to a straw colour. This may occur after buds have hardened. The spray should be directed to avoid leaders, especially in seasons when lammas growth occurs. Other species should not sprayed overall when using the rates applicable for heather control.

Product rates

On mineral soils:	6 litres of product per hectare.
On peaty soil:	4 litres of product per hectare.

Methods of application

1. *Pre-plant (overall or band application)*

Tractor mounted equipment

Boom sprayer at LV or MV.

Ulvaforest low speed rotary atomiser at VLV.

Handheld equipment

Knapsack sprayer at MV or LV.
Herbi low speed rotary atomiser at VLV.
Forestry spot gun at LV.

2. *Post-plant (directed application)*

Handheld equipment

Knapsack sprayer at MV or LV.
Knapsack sprayer with 'VLV' nozzle at LV.
Herbi low speed rotary atomiser at VLV.
Weedwiper for direct application uncalibrated – see product rate section.

Refer to Section 11 for details of applicator and correct calibration.

Timing of application

Late August to end of September after new growth on crop trees has hardened (see Crop tolerance). This is also the optimum time for the control of heather pre-planting.

Additional information

1. **Weed control**

a. Glyphosate applications in August and September to control heather will be too late to lessen the effect of weed competition in the current season.

b. Glyphosate is most effective when relative humidity is high and the air is warm (e.g. 15°C+).

c. Reduced weed control may result when weeds are under stress, for example from factors such as drought.

d. Heavy rainfall within 24 hours of application may reduce foliar absorption. The addition of Mixture B at 2% of final spray volume will improve reliability in these circumstances. The addition of Mixture B will reduce crop tolerance and it must only be used pre-plant or post-plant as a directed spray.

e. Diluted product may denature after 2 to 3 days if clean water is not used. Where possible use tap water as the diluent and only mix sufficient for the day's programme.

2. **Protective clothing**

Read the product label for protective clothing and equipment requirements and check there are no items required in addition to the Forestry Commission recommendation in Section 10.

3. **Special precautions**

a. Unlike the additional glyphosate products listed at the end of this Section, Roundup Pro Biactive and Roundup Biactive DO NOT require a warning on the product label about harmful effects to aquatic life, or about irritant effects to operators. However, the addition of Mixture B makes any application potentially harmful to aquatic life, and a potential irritant to operators.

b. Do not contaminate ponds, watercourses or ditches with the chemical or used containers.

c. While bees are unaffected by glyphosate, heather flowering will probably be affected by applications earlier than August. Beekeepers should be advised not to site hives on areas to be sprayed.

The label on the herbicide container has been designed for your protection – ALWAYS READ THE INSTRUCTIONS ON THE LABEL.

Additional glyphosate products with full forestry approval

Barclay Gallup	360 g/litre glyphosate (Barclay)
Barclay Gallup Amenity	360 g/litre glyphosate (Barclay)
Clayton Glyphosate	360 g/litre glyphosate (Clayton)
Clayton Swath	360 g/litre glyphosate (Clayton)
Glyphogan	360 g/litre glyphosate (PBI)
Glyphosate-360	360 g/litre glyphosate (Top Farm)
Helosate	360 g/litre glyphosate (Helm)
Hilite	144 g/litre glyphosate (Nomix-Chipman) – CDA formulation
Outlaw	360 g/litre glyphosate (Barclay)
Portman Glyphosate 360	360 g/litre glyphosate (Portman)
Roundup	360 g/litre glyphosate (Monsanto)
Roundup	360 g/litre glyphosate (Schering/Agro Evo)
Roundup Biactive Dry	42.6% w/w glyphosate (Monsanto)
Stacato	360 g/litre glyphosate (Unicrop)
Stefes Glyphosate	360 g/litre glyphosate (Stefes)
Stefes Kickdown 2	360 g/litre glyphosate (Stefes)
Stetson	360 g/litre glyphosate (Monsanto)
Stirrup	144 g/litre glyphosate (Nomix-Chipman) – CDA formulation

These products may have different conditions of use regarding operator and environmental safety – refer to the product label for further guidance.

6.4 Imazapyr

Product

Arsenal 50F 50 g/litre imazapyr
(Cyanamid, marketed by Nomix-Chipman)

Description

A translocated and residual herbicide which will control a wide range of established annual and perennial grass and herbaceous broadleaved weeds, and will provide long term residual control of germinating seedlings. Severe check of heather occurs but information on long term control has not been fully evaluated.

Imazapyr is absorbed through roots and foliage, then rapidly translocated throughout the plant. Treated roots stop growing soon after application, but chlorosis and tissue necrosis may not be apparent until up to 2 weeks after application – complete kill may take several weeks.

Crop tolerance

Imazapyr should only be used as a pre-plant treatment. The only tolerant species according to the product label are SS, LP and CP. Limited FC trials suggest JL, DF and NS may also be tolerant. The trees must be at least 2 year old nursery stock, and 5 months must elapse between treatment and subsequent planting. All other species may be damaged.

Product rate

15 litres per hectare.

Methods of application

Pre-plant (overall or band)

Tractor mounted equipment

Boom sprayer at LV–MV (200 l/ha).
Ulvaforest low speed rotary atomiser at VLV.

Handheld equipment

Knapsack sprayer at LV–MV (200 l/ha).
Forestry spot gun at LV.

Refer to Section 11 for details of applicators and correct calibration.

Timing of application

Apply from July–October, when weeds are actively growing.

Additional information

1. **Weed control**

a. Residual weed control will be reduced if cultivation takes place after spraying. Ideally, a site should be sprayed after cultivation has occurred, and when any weeds present have started to re-grow. When this is not possible, a period of at least a week should elapse after treatment, prior to cultivation.

b. Do not use imazapyr on soils that may later be used for growing desirable plants, except for the coniferous species SS, CP and LP.

c. Do not use on conifer nursery beds.

d. Do not use near desirable trees or shrubs, nor in areas into which their roots may extend or in locations where the chemical may be washed or moved into contact with their roots.

e. Do not apply in windy conditions, or using high pressures producing a fine spray liable to drift.

f. The length of residual weed control given has not been fully evaluated, but indications are that in some circumstances weed suppression may still occur up to two seasons after planting. However, this may vary from site to site, and there are no manufacturer's recommendations on this subject at present.

g. Heather flowering will be affected by applications – beekeepers should be advised not to site hives on areas to be sprayed.

h. Results from trials suggest that planted trees are unlikely to be damaged from root contact with treated vegetation, or from leachates from dead vegetation that is subsequently cultivated. However, imazapyr may be very mobile in soil.

2. Protective clothing

Read the product label for protective clothing and equipment requirements, and check there are no items required in addition to the Forestry Commission recommendations in Section 11.

3. Special precautions

a. Irritating to eyes – avoid contact.

b. Do not contaminate ponds, watercourses or ditches with the chemical or used containers.

The label on the herbicide container has been designed for your protection – ALWAYS READ THE INSTRUCTIONS ON THE LABEL.

7 Woody weeds

7.1 General

This group of weeds contains a wide range of species including brambles, climbers, gorse and broom, shrubs and all types of tree. Species such as birch in some circumstances are part of the timber crop, but in others are unwanted and may threaten the resource. Such situations require a precise definition of management objectives and constraints, such as the present and future amenity effects of broadleaved components of a stand.

The biological similarity of woody weeds to crop trees can make selective chemical control difficult or impossible. As perennial woody plants, many with the ability to coppice strongly, they present a range of weeding situations requiring a variety of control methods.

Different ways of treating unwanted regeneration may include:

a. Overall, non-selective pre-plant spray – no crop tolerance problems.
b. Selective pre-plant spray – no crop tolerance problems.
c. Selective post-plant spray – severe crop tolerance problems. In most cases a directed spray needs to be used. Even then there may be problems of translocation to untreated regeneration or crop trees.

Herbicides with forestry approval for this use are:

Ammonium sulphamate – stem treatment, cut stump;
2,4-D – foliar treatment;
2,4-D/dicamba/triclopyr – foliar treatment, cut stump;
Fosamine ammonium – foliar treatment;
Glyphosate – foliar treatment, stem treatment, cut stump.
Imazapyr – foliar treatment.
Triclopyr – foliar treatment, stem treatment, cut stump.

Fosamine ammonium and 2,4-D are unlikely to give adequate control of evergreen species. 2,4-D/dicamba/triclopyr and triclopyr will give good control of most woody and herbaceous species, but will have little effect on any grass weeds on site. Triclopyr is particularly effective against gorse and broom and again will have little effect on any grass present. Glyphosate and imazapyr will control most weeds. All of the herbicides can cause damage to crop plants if they are used inappropriately.

Selective chemical respacing of broadleaved natural regeneration may be best achieved from using directed sprays of 2,4-D, fosamine ammonium, glyphosate or triclopyr. Glyphosate or triclopyr should be used on coniferous species. In both cases, cut stump treatment may be necessary to prevent re-growth. The other herbicides in this section will control natural regeneration, but due to their soil action selective control may be difficult.

When treating species with edible fruit, for example brambles, it is important to erect notices warning visitors not to collect fruit from sprayed areas.

7.2 Types of treatment

Control methods fall into three groups:

Foliar treatment

This is normally the preferred method of application wherever vegetation is in leaf and the foliage is accessible to be sprayed. Weeds must however be small enough to allow an effective dose of herbicide to be applied to the foliage. When treating small weeds or regeneration less than 1 metre tall it is practicable to consider application rates on a per treated hectare basis. But when weeds are tall and are growing in clumps or bushes it is more practicable to consider applying a herbicide solution of a given strength to cover all foliage to wetness. Tall stems which need to be cut and cannot for some reason be stump treated may be allowed to regrow for 1 or 2 years before spraying regrowth. Table 6 gives an indication of the susceptibilities of the various species to the available herbicides.

Stem treatments

The techniques of frill girdling and stem injection are appropriate for the treatment of woody growth that is too tall to allow the foliage to be sprayed effectively and where the resulting dead standing stems can be accepted or conveniently removed. Costs can be high if there is a high stocking of unwanted woody growth.

Cut stump treatment

This method is used for control of coppice regrowth after felling (of crop trees or scrub) and avoids the problem of unsightly dead stems remaining standing on the site. As with stem treatments, the cost per treated hectare can be high if there is a high stocking of unwanted woody growth.

Table 6 Product rates for woody weeds susceptible to foliar application

Species	Herbicide rate: (l/ha) glyphosate*1	(l/ha) triclopyr	(l/ha) 2,4-D	(% of product in water) 2,4-D/dicamba/triclopyr	(l/ha) Fosamine ammonium	(l/ha) imazapyr*5
Acer spp. (sycamore, maple)	3	4	–	2%*3	10 MS	5–10
Alnus spp. (alder)	–	4	8–13 MS	–	10	5–10
Betula spp. (birch)	2	4	8–13 MS	2%	10	5–10
Castanea sativa (chestnut)	–	6	–	2%	–	–
Cornus sanguinea (dogwood)	–	4	–	–	10 MS	10–15
Corylus avellana (hazel)	3	6	R	2%*3	10	5–10
Crataegis monogyna (hawthorn)	3	MR*2	R	–	10	10–15
Cytisus scoparius (broom)	MR	2	–	2%	R	10
Fagus sylvatica (beech)	2	6	–	2%	10	5–10
Fraxinus excelsior (ash)	3	8	MR	2%	10	5–10
Ligustrum vulgare (privet)	2	6	–	–	–	10–15
Polygonium japonicum (Japanese knotweed)*4	6	–	–	6.5%	–	–
Populus spp. (aspen poplar)	2	4	–	–	10	5–10
Prunus laurocerasus (laurel)	8–10	8	–	–	–	10–15
Prunus spinosa (blackthorn)	2	4	–	–	10	10–15
Quercus spp. (oak)	3	8	R	2%	10	5–10
Rhamnus cathartica (buckthorn)	–	6	–	2%	10	5–10
Rhododendron ponticum (rhododendron)	8–10	8	R	7.5%*3	R	4–15*6
Rosa canina (wild rose)	2	2	–	–	10	10
Rubus spp. (bramble)	3	2	MR	1%	10	10–15
Salix spp. (willow)	3	6	8–13	–	10 MS	5–10

Table 6

Product rates for woody weeds susceptible to foliar application

Species	Herbicide rate: (l/ha) glyphosate*1	(l/ha) triclopyr	(l/ha) 2,4-D	(% of product in water) 2,4-D/dicamba/triclopyr	(l/ha) Fosamine ammonium	(l/ha) imazapyr*5
Sambucus nigra (elder)	2	4	8–13 MS	–	10	5–10
Sorbus aucuparia (rowan)	2	6	8–13 MS	–	10 MS	–
Ulex spp. (gorse)	R/MR*7	2	MR	2%	R	15
Ulmus procera (elm)	2	6	MR	–	10	5–10
Viburnum opulus (gelder rose)	5	4	–	–	–	–

Notes:

MR Moderately resistant.

MS Moderately susceptible.

R Resistant.

1% 1 ml of product per litre water.

– Not tested/No manufacturer's information.

*1 For pre-plant treatments, particularly in areas with dense growth or moderately resistant species the rate should be increased to 5 litres per hectare to ensure effective control (retain the 8–10 l/ha rate for laurel/rhododendron).

*2 Susceptible to stump treatment see Section 7.5.4.

*3 May require more than one application for complete control.

*4 Use only glyphosate to control Japanese knotweed near water.

*5 With the exception of *Ulex* spp., these are not UK approved label recommendations. The data is based on US and European trials – the manufacturer will not guarantee control in UK conditions.

*6 Results of FC trials – use the higher rate if only partial coverage of the bush is possible.

*7 Seedlings susceptible.

7

7.3 Foliar treatment

7.3.1 2,4-D

Product

Dicotox Extra 400 g/litre 2,4-D (Rhone Poulenc)

Description

2,4-D is a plant growth regulating herbicide to which many herbaceous and woody broadleaved species are susceptible. It is absorbed mainly through aerial parts of the plant, but also through the roots.

One month should elapse between treatment and subsequent planting.

Crop tolerance

In general, overall sprays should not be made, but directed to avoid contact with foliage, immature bark or leaders.

SS, NS and OMS are moderately tolerant of overall sprays provided leader growth has hardened. Date of hardening can be variable – it may occur as early as the end of July or may be delayed until the end of October in some locations and seasons.

SP, CP, DF, WRC, GF and NF are rather less tolerant and spray should be directed away from foliage, particularly leading shoots.

LP, WH and larches are sensitive.

Broadleaves are very sensitive to 2,4-D.

Hot weather at the time of spraying may increase crop damage.

Product rate

In young conifer plantations (>1 m in height) – 8 litres per treated hectare.
In established trees – 13 litres per treated hectare
– see Table 6.

Methods of application

Pre-plant (overall or band) or post-plant (directed application)

Tractor mounted equipment

Boom sprayer at MV.

Handheld equipment

Knapsack sprayer at MV.

Forestry spot gun at LV.

Refer to Section 11 for details of applicators and current calibration.

Timing of application

Apply from May to August after crop has hardened off. Applications earlier than mid-July should only be applied as a directed spray so as to avoid serious crop damage.

Additional information

1. **Weed control**

a. To reduce the risk of crop damage trees should be 1 metre tall to minimise contact of spray mixture and leading shoots.

b. Volatilisation can cause serious damage to neighbouring crops – if possible avoid applications in hot weather.

c. Broadleaved plants are very sensitive to 2,4-D especially in warm weather, avoid spraying when wind would cause drift and damage to neighbouring crops or trees.

2. **Protective clothing**

Read the product label for protective clothing and equipment requirements and check there are no items required in addition to the Forestry Commission recommendations in Section 10.

3. **Special precautions**

a. Special precautions are required in water catchment areas to avoid water taint – see Section 3.6.

b. If possible 2,4-D should not be applied in areas visited regularly by the public in significant numbers, and where edible fruits or plants are likely to be exposed to spray. If, however, spraying does take place in such areas appropriate warning notices must be erected. Such signs should remain in position as long as treated fruit appear wholesome.

c. Harmful to fish; do not contaminate ponds, watercourses or ditches with the chemical or used container.

The label on the herbicide container has been designed for your protection – ALWAYS READ THE INSTRUCTIONS ON THE LABEL.

7

7.3.2 2,4-D/Dicamba/Triclopyr

Product

Broadshot 200:85:65 g/litre 2,4-D/dicamba/triclopyr (Cyanamid)

Description

A translocated herbicide which controls a wide range of annual and perennial herbaceous weeds, as well as woody weeds. Most grasses are resistant, although some may be suppressed.

Crop tolerance

Most conifers are tolerant when fully dormant but contact from spray with crop foliage or rooting zone should be avoided.

All broadleaves are susceptible to any contact.

Product rate

Apply as a 1–7.5% solution of product in water, depending on species – see Table 6. Maximum product rate to prevent potential environmental contamination is 5 l/ha.

For direct application with the weedwiper, mix 1 part product with 8 parts water. Red dye (available from Hortichem Ltd.) may be added to identify treated areas.

Methods of application

Pre-plant (band or overall) or post-plant (directed spray)

Tractor mounted equipment

Boom sprayer at LV or MV (150–400 l/ha).
Use the higher volume rate for larger, denser weed populations.

Handheld equipment

Weedwiper for direct application – uncalibrated, refer to product rate section.
Knapsack sprayer at LV or MV.
Refer to Section 11 for details of applicators and correct calibration.

Timing of application

Apply July–September, when spring growth has slowed, but before signs of senescence.

Additional information

1. **Weed control**

a. Do not apply when rain is imminent, or in periods of very hot or cold weather.

b. Volatilisation can cause serious damage to neighbouring crops – if possible avoid applications in hot weather.

c. Some woody species may require more than one application – see Table 6.

d. After spraying woody vegetation, 3 months delay should be allowed before replanting.

2. **Protective clothing**

Read the product label for protective clothing and equipment requirements, and check there are no items required in addition to the Forestry Commission recommendations in Section 10.

3. **Special precautions**

a. Irritating to eyes and skin – avoid contact. Harmful if swallowed.

b. The herbicide can be dangerous to fish – do not contaminate ponds, watercourses or ditches with the chemical or used containers.

The label on the herbicide container has been designed for your protection – ALWAYS READ THE INSTRUCTIONS ON THE LABEL.

7.3.3 Fosamine ammonium

Product

Krenite 480 g/litre fosamine ammonium (Du Pont)

This product is approved at the time of writing, but manufacture is likely to cease in the near future.

Description

Fosamine ammonium is a non-translocated foliar acting herbicide for deciduous woody weed control, in non-crop areas in forests. After application little effect is visible until the following spring when bud and shoot development is severely limited, which is followed by death of the treated plant. Grasses, herbaceous vegetation and evergreen woody species are unlikely to be affected, but it is advisable to test a small area before large scale treatment.

Crop tolerance

Fosamine ammonium should only be used in non-crop areas. Herbicide drift will result in severe damage to all actively growing trees, and may damage broadleaved species, larch and pine, even if they are dormant.

Product rate

Apply at 10 litres per treated area. Ensure the spray solution is of a concentration not less than 1.5% of the product in water. Combine with an approved non-ionic wetting agent, such as Mixture B (2% by volume). See Table 6.

Methods of application

Pre-plant on non-crop areas (overall, band or spot)

Tractor mounted equipment

Boom sprayer at MV.

Handheld equipment

Knapsack sprayer at LV–MV (at least 100 l/ha).

Apply to wet all foliage.

Refer to Section 11 for details of applicators and correct calibration.

Timing

Apply in August–October, before autumn discolouration of foliage.

Additional information

1. **Weed control**

a. Do not apply to wet foliage. Rainfall within 24 hours of application will reduce performance.

b. Trees and shrubs close to water may be treated, but do not spray directly on to watercourses.

c. Thorough coverage of weed species is necessary for effective treatment.

d. Fosamine ammonium is rapidly broken down in the soil.

2. **Protective clothing**

Read the product label for protective clothing and equipment requirements, and check there are no items required in addition to the Forestry Commission recommendations in Section 10.

3. **Special precautions**

Do not contaminate ponds, watercourses or ditches with the chemical or the used container.

The label on the container has been designed for your protection – ALWAYS READ THE INSTRUCTIONS ON THE LABEL.

7.3.4 Glyphosate

Products

Roundup Biactive 360 g/litre glyphosate (Monsanto)
Roundup Pro Biactive 360 g/litre glyphosate (Monsanto)

Description

A translocated herbicide taken up by the foliage and conveyed to the roots. It causes chlorosis and eventual death of leaves and kills roots and shoots.

Glyphosate controls a wide range of weeds including grasses, herbaceous broadleaved weeds, bracken, heather and woody weeds. When applied late in the growing season, the main effect is obtained in the following year when roots die and suckering is prevented.

On contact with the soil, glyphosate is quickly inactivated. Planting can be carried out 7 days after treatment, and a minimum of 7 days should be allowed before cultivation and the breaking up of rhizomes and roots.

Crop tolerance

SS, NS, SP, CP, LP, WRC and LC will tolerate overall sprays provided leader growth has hardened. Hardening can occur as early as the end of July or may be delayed until the end of October in some locations and seasons. Hardening may be indicated when the leader changes from bright green to a straw colour. This may occur after buds have hardened. To avoid damage to lammas growth, herbicide sprays must be directed away from leaders. During the active growing season trees must be guarded or the spray placed to avoid contact with the crop.

DF and NF: as above but much more sensitive.

Broadleaved trees, larch and other conifers will not tolerate overall applications: always use a guard, a weedwiper or a directed spray to avoid contact with the foliage and immature bark of crop trees.

Product rate

Apply 2.0–5.0 litres (see Table 6) of product per treated hectare, diluted in water and sprayed to wet all foliage.

For direct application with the weedwiper, use 1 part product in 2 parts water. Red dye (available from Hortichem Ltd.) can be used to mark treated areas if required.

Methods of application

Pre-plant (overall or band application) or post-plant (overall, band or directed application)

Tractor mounted equipment

Boom sprayer at LV or MV.

Ulvaforest low speed rotary atomiser at VLV.

Handheld equipment

Knapsack sprayer at MV.

Knapsack sprayer at LV.

ULVA high speed rotary atomiser at VLV. For full effect, dilute the herbicide with at least 5 times the volume of water.

Weedwiper for direct application.

Forestry spot gun at LV.

Refer to Section 11 for details of applicator and correct calibration

Timing of application

June to August inclusive but after new growth on crop trees has hardened (see Crop tolerance). This is also the optimum time for pre-planting sprays of woody regrowth.

Additional information

1. Weed control

a. Glyphosate applied later than June will be too late to lessen the effect of weed competition in the current season.

b. Glyphosate is more effective when relative humidity is high and the air is warm (e.g. 15°C+).

c. Reduced weed control may result when weeds are under stress, e.g. frost or drought.

d. Heavy rainfall within 24 hours of application may reduce the herbicide's effectiveness by preventing sufficient foliar absorption. The addition of Mixture B at 2% of final spray volume will improve reliability in these circumstances. The addition of Mixture B will reduce crop tolerance and it must only be used pre-plant or post-plant as a directed spray.

e. Diluted product may denature after 2 to 3 days if clean water is not used. Where possible use tap water as the diluent and only mix sufficient for the day's programme.

2. **Protective clothing**
   Read the product label for protective clothing and equipment requirements and check there are not items required in addition to the Forestry Commission recommendations in Section 10.
3. **Special precautions**

a. Unlike the additional glyphosate products listed at the end of this Section, Roundup Pro Biactive and Roundup Biactive DO NOT require a warning on the product label about harmful effects to aquatic life, or about irritant effects to operators. However, the addition of Mixture B makes any application potentially harmful to aquatic life, and a potential irritant to operators.
b. Do not contaminate ponds, watercourses or ditches with the chemical or used containers.

   The label on the herbicide container has been designed for your protection – ALWAYS READ THE INSTRUCTIONS ON THE LABEL.

Additional glyphosate products with full forestry approval

Barclay Gallup	360 g/litre glyphosate (Barclay)
Barclay Gallup Amenity	360 g/litre glyphosate (Barclay)
Clayton Glyphosate	360 g/litre glyphosate (Clayton)
Clayton Swath	360 g/litre glyphosate (Clayton)
Glyphogan	360 g/litre glyphosate (PBI)
Glyphosate-360	360 g/litre glyphosate (Top Farm)
Helosate	360 g/litre glyphosate (Helm)
Hilite	144 g/litre glyphosate (Nomix-Chipman) – CDA formulation
Outlaw	360 g/litre glyphosate (Barclay)
Portman Glyphosate 360	360 g/litre glyphosate (Portman)
Roundup	360 g/litre glyphosate (Monsanto)
Roundup	360 g/litre glyphosate (Schering/Agro Evo)
Roundup Biactive Dry	42.6% w/w glyphosate (Monsanto)
Stacato	360 g/litre glyphosate (Unicrop)
Stefes Glyphosate	360 g/litre glyphosate (Stefes)
Stefes Kickdown 2	360 g/litre glyphosate (Stefes)
Stetson	360 g/litre glyphosate (Monsanto)
Stirrup	144 g/litre glyphosate (Nomix-Chipman) – CDA formulation

These products may have different conditions of use regarding operator and environmental safety – refer to the product label for further guidance.

7.3.5 Imazapyr

Product

Arsenal 50F   50 g/litre imazapyr
(Cyanamid, marketed by Nomix-Chipman)

Description

A translocated and residual herbicide which will control a wide range of established annual and perennial grass and herbaceous broadleaved weeds, and will provide long term residual control of germinating seedlings. Control of woody weeds is likely to be good, although there are no manufacturer's label recommendations at present. Imazapyr is absorbed through roots and foliage, then rapidly translocated throughout the plant. Treated plants stop growing soon after application, but chlorosis and tissue necrosis may not be apparent until up to 2 weeks after application – complete kill may take several weeks.

Crop tolerance

Imazapyr should only be used as a pre-plant treatment. The only tolerant species according to the product label are SS, LP and CP. Limited FC trials suggest JL, DF and NS may also be tolerant. The trees must be at least 2 year old nursery stock, and 5 months must elapse between treatment and subsequent planting. All other species will be damaged.

Product rate

There are no manufacturer's label recommendations for woody weed control, but the highest rate of 15 litres per treated hectare is likely to severely check woody weeds, and provide a measure of long term control – see Table 6.

Methods of application

Pre-plant (overall or band)

Tractor mounted equipment

Boom sprayer at LV–MV (200 l/ha).

Ulvaforest low speed rotary atomiser at VLV.

Handheld equipment

Knapsack sprayer at LV–MV (200 l/ha).

Forestry spot gun at LV.

Refer to Section 11 for details of applicators and correct calibration.

 ALWAYS READ THE PRODUCT LABEL

Timing of application

Apply from July–October, when weeds are actively growing.

Additional information

1. **Weed control**

a. Residual weed control will be reduced if cultivation takes place after spraying. Ideally, a site should be sprayed after cultivation has occurred, and when any weeds present have started to regrow. When this is not possible, a period of at least a week should elapse after treatment, prior to cultivation.

b. Do not use imazapyr on soils that may later be used for growing desirable plants, except for the coniferous species SS, CP and LP.

c. Do not use on conifer nursery beds.

d. Do not use near desirable trees or shrubs, nor in areas into which their roots may extend or in locations where the chemical may be washed or moved into contact with their roots.

e. Do not apply in windy conditions, or using high pressures producing a fine spray liable to drift.

f. The length of residual weed control given has not been fully evaluated, but indications are that in some circumstances weed suppression may still occur up to two seasons after planting. However, this may vary from site to site, and there are no manufacturer's recommendations at present.

g. Results of trials suggest that planted trees are unlikely to be damaged from root contact with treated vegetation, or from leachates from dead vegetation that is subsequently cultivated. However, imazapyr may be very mobile in soil.

2. **Protective clothing**

Read the product label for protective clothing and equipment requirements, and check there are no items required in addition to the Forestry Commission recommendations in Section 10.

3. **Special precautions**

a. Irritating to eyes – avoid contact.

b. Do not contaminate ponds, watercourses or ditches with the chemical or used containers.

The label on the herbicide container has been designed for your protection – ALWAYS READ THE INSTRUCTIONS ON THE LABEL.

7.3.6 Triclopyr

Products

Timbrel	480 g/litre triclopyr (DowElanco)
Garlon 4	480 g/litre triclopyr (DowElanco)
Chipman Garlon 4	480 g/litre triclopyr (Nomix-Chipman)

Description

A plant growth regulating herbicide which is rapidly absorbed, mainly through the foliage, but also by roots and stems. Once inside the plant it is translocated. Triclopyr is effective against many woody weeds (see Table 6), and is particularly effective against gorse and broom.

Grasses sometimes show some yellowing following spraying operations but this is quickly outgrown. Most herbaceous vegetation will be killed.

In the soil triclopyr is broken down fairly rapidly by microbial action. Planting should be deferred for at least 6 weeks after application.

Crop tolerance

SS and NS will tolerate overall sprays up to 4 l/ha provided leader growth has hardened. Hardening can occur as early as the end of July or may be delayed until October in some locations and seasons. To avoid damage to lammas growth, herbicide sprays must be directed away from leaders. During the active growing season trees must be guarded or the spray placed to avoid contact with the crop.

DF and pines are rather more sensitive than spruces with occasional leader damage if sprayed overall.

Larch, other conifers and broadleaves are severely damaged by overall sprays at any time of year. Late September application will be tolerated if applied with great care to avoid the foliage of crop trees.

Product rate

Apply 2.0–8.0 litres (see Table 6) of product per treated hectare, diluted in water and sprayed to wetness.

Methods of application

1. *Pre-plant (overall or band application)*

Tractor mounted equipment

Tractor mounted boom sprayer at MV.

Handheld equipment

Knapsack sprayer at MV.

2. *Post-plant (directed, overall or band application)*

Handheld equipment

Knapsack sprayer (with guard if required) at MV.

Forestry spot gun.

Refer to Section 11 for details of applicator and correct calibration.

Timing of application

The timing of overall applications will be determined by the growth stage of the crop trees (see Crop tolerance), but to be effective the operation must take place after crop trees have hardened but before the target species show signs of senescence.

Directed applications can take place at any time between June–September provided the target species shows no signs of senescence and adequate precautions are taken to protect the crop species.

Applications to control gorse will have greatest effect from April–October.

Additional information

1. **Weed control**

a. Volatilisation can cause serious damage to neighbouring crops – if possible avoid applications in hot weather.

b. Rainfall within 2 hours of application may reduce the herbicide's effectiveness by preventing sufficient foliar absorption.

c. A minimum interval of 6 weeks is required between application of triclopyr and planting.

2. **Protective clothing**

Read the product label for protective clothing and equipment requirements and check there are no items required in addition to the Forestry Commission recommendations in Section 10.

3. **Special precautions**

a. Triclopyr is harmful if in contact with skin and irritating to eyes – avoid contact.

b. Triclopyr should not be applied via controlled droplet applicators.

c. Triclopyr is dangerous to fish: do not contaminate ponds, watercourses or ditches with the concentrate, spray drift or used containers.

d. Do not apply with tractor mounted boom sprayers within 250 m of ponds, lakes or watercourses.

The label on the herbicide container has been designed for your protection – ALWAYS READ THE INSTRUCTIONS ON THE LABEL.

7.4 Stem treatment

7.4.1 Ammonium sulphamate

Products

Amcide soluble crystals	100% ammonium sulphamate (Battle, Hayward and Bower)
Root-out soluble crystals	100% ammonium sulphamate (Dax)

Description

A highly soluble translocated, contact and soil-acting herbicide which is absorbed through leaves, roots and exposed live tissue surfaces. It is effective against most woody species including the more resistant species such as rhododendron, hawthorn and ash. It corrodes metals and alloys including copper, brass, mild steel and galvanised iron.

Breakdown in the soil can take up to 12 weeks during which time it retains its herbicidal properties. After breakdown only natural elements remain – there are no complex organic residues. Three months should elapse between treatment and subsequent planting.

Crop tolerance

All crop species are severely damaged or killed by direct application of ammonium sulphamate or by absorption via the roots.

Post-planting frill-girdling or notching can be carried out safely if great care is taken to avoid unnecessary spillage or overflow reaching the ground. Pre-planting treatment is preferable.

Product rates and methods of application

Frill girdling

A frill is cut round each stem by overlapping downward strokes of a light axe or billhook. Exposed live tissue is sprayed to run-off with a 40% solution of ammonium sulphamate (0.4 kg crystals per 1 litre of water) using a plastic watering can or a Technoma T18P semi-pressurised knapsack sprayer (see Section 11.5.4).

Notching

Individual notches are cut round the stem by using downward axe strokes penetrating the live cambial tissue. Notches should be at least 3 cm long and should not be further than 10 cm apart, edge to edge. 15 g of dry crystals are placed in each notch. They should be sited between ground level and 1 metre high.

Timing of application

Best results are obtained from applications made during the growing season but ammonium sulphamate can be applied at any time between April and September.

Additional information

1. **Weed control**

a. Ammonium sulphamate is best applied in dry weather so that the spray solution or the crystals are not washed out of the stem frill or notch.

b. Solutions of ammonium sulphamate in water should be freshly prepared on each day of use.

2. **Protective clothing**

Read the product label for protective clothing and equipment requirements and check there are no items required in addition to the Forestry Commission recommendations in Section 10.

3. **Special precautions**

Ammonium sulphamate is harmful to fish; do not contaminate ponds, watercourses or ditches with the chemical or used container.

The label on the herbicide container has been designed for your protection – ALWAYS READ THE INSTRUCTIONS ON THE LABEL.

7.4.2 Glyphosate

Products

Roundup Biactive	360 g/litre glyphosate (Monsanto)
Roundup Pro Biactive	360 g/litre glyphosate (Monsanto)

Description

A translocated herbicide normally applied to and taken up by the foliage but which is also effective as a stem injection treatment.

It causes chlorosis and eventual death of leaves and kills roots and shoots.

Stem injection with glyphosate controls all the major broadleaved woody weed species and is also highly effective in killing individual stems of SS and other conifers, e.g. in chemical thinning.

Crop tolerance

There is little evidence of translocation across root grafts to untreated trees ('flashback'). In general, unwanted stems can be safely treated by this method among any crop species.

For foliar crop tolerance see Section 7.3.4.

For treatment of stems with a cleaning saw and application of glyphosate via a cleaning saw attachment to the cut stump, see Section 7.5.3.

Product rate

Apply 2 ml of neat glyphosate per 10 cm stem diameter.

For trees with a diameter greater than 10 cm, separate incisions evenly spaced around the stem should be made for each 10 cm stem diameter.

Method of application

Application should be made using downward axe strokes to penetrate the live cambial tissue and then applying glyphosate into the notches using a Forestry spot gun;

For trees up to 10 cm diameter: cut a single notch on one side of the stem.

For larger trees a second cut should be made on the opposite side of the stem.

Refer to Section 11 for details of applicators.

Timing of application

At any time of year except during the period of maximum sap flow in spring. This usually occurs during March–May.

Additional information

1. **Weed control**

   Glyphosate applied later than June will be too late to lessen the effect of weed competition in the current season.

2. **Protective clothing**

   Read the product label for protective clothing and equipment requirements and check there are no items required in addition to the Forestry Commission recommendations in Section 10.

3. **Special precautions**

a. Unlike the additional glyphosate products listed at the end of this Section, Roundup Pro Biactive and Roundup Biactive DO NOT require a warning on the product label about harmful effects to aquatic life, or about irritant effects to operators. However, the addition of Mixture B makes any application potentially harmful to aquatic life, and a potential irritant to operators.

b. Do not contaminate ponds, watercourses or ditches with the chemical or used containers.

The label on the herbicide container has been designed for your protection – ALWAYS READ THE INSTRUCTIONS ON THE LABEL.

Additional glyphosate products with full forestry approval

Barclay Gallup	360 g/litre glyphosate (Barclay)
Barclay Gallup Amenity	360 g/litre glyphosate (Barclay)
Clayton Glyphosate	360 g/litre glyphosate (Clayton)
Clayton Swath	360 g/litre glyphosate (Clayton)
Glyphogan	360 g/litre glyphosate (PBI)
Glyphosate-360	360 g/litre glyphosate (Top Farm)
Helosate	360 g/litre glyphosate (Helm)
Hilite	144 g/litre glyphosate (Nomix-Chipman) – CDA formulation
Outlaw	360 g/litre glyphosate (Barclay)
Portman Glyphosate	360 g/litre glyphosate (Portman)
Roundup	360 g/litre glyphosate (Monsanto)
Roundup	360 g/litre glyphosate (Schering/Agro Evo)
Roundup Biactive Dry	42.6% w/w glyphosate (Monsanto)

Stacato	360 g/litre glyphosate (Unicrop)
Stefes Glyphosate	360 g/litre glyphosate (Stefes)
Stefes Kickdown 2	360 g/litre glyphosate (Stefes)
Stetson	360 g/litre glyphosate (Monsanto)
Stirrup	144 g/litre glyphosate (Nomix-Chipman) – CDA formulation

These products may have different conditions of use regarding operator and environmental safety – refer to the product label for further guidelines.

7.4.3 Triclopyr

Products

Timbrel	480 g/litre triclopyr (DowElanco)
Garlon 4	480 g/litre triclopyr (DowElanco, marketed by Nomix-Chipman)
Chipman Garlon 4	480 g/litre triclopyr (Nomix-Chipman)

Description

A plant growth regulating herbicide which is rapidly absorbed mainly through the foliage but also by roots and stems; once inside the plant it is readily translocated. Triclopyr is effective against a wide range of weed species but hawthorn is relatively resistant. It is particularly effective against gorse and broom. Grasses sometimes show some yellowing following spraying operations but this is quickly outgrown. Herbaceous weeds will be controlled.

In the soil Timbrel is broken down fairly rapidly by microbial action. Planting should be deferred for at least 6 weeks after application.

Crop tolerance

There is no evidence of translocation across root grafts to untreated trees ('flashback'). Unwanted stems can be safely treated by this method among any crop species.

For foliar crop tolerance see Section 7.3.6.

Product rate

Apply 2 ml undiluted or 4 ml diluted (1:1 with water) per 5 cm stem diameter.

For trees with trunks larger than 5 cm diameter, separate incisions evenly spaced around the stem should be made for each 5 cm stem diameter.

Method of application

Application should be made using downward axe strokes to penetrate the live cambial tissue and then applying Timbrel into the notches using a Forestry spot gun.

Refer to Section 11 for details of applicators.

Timing of application

Applications may be made at any time of year but best results follow summer or autumn treatments.

Additional information

1. **Weed control**
   Volatilisation can cause serious damage to neighbouring crops – if possible avoid applications in hot weather.
2. **Protective clothing**
   Read the product label for protective clothing and equipment requirements and check there are no items required in addition to the Forestry Commission recommendations in Section 10.
3. **Special precautions**

a. Triclopyr is harmful if in contact with skin and irritating to eyes – avoid contact.

b. Triclopyr is dangerous to fish – do not contaminate ponds, watercourses or ditches with the concentrate, spray drift or used containers.

The label on the herbicide container has been designed for your protection – ALWAYS READ THE INSTRUCTIONS ON THE LABEL.

7.5 Cut stump treatment

7.5.1 Ammonium sulphamate

Products

Amcide soluble crystals	100% ammonium sulphamate (Battle, Hayward and Bower)
Root-out soluble crystals	100% ammonium sulphamate (Dax)

Description

A highly soluble translocated, contact and soil-acting herbicide which is absorbed through leaves, roots and exposed live tissue surfaces. It is effective against most woody species including the more resistant species such as rhododendron, hawthorn and ash. It corrodes metals and alloys including copper, brass, mild steel and galvanised iron.

Breakdown in the soil can take up to 12 weeks during which time it retains its herbicidal properties. Three months should elapse between treatment and subsequent planting.

After breakdown only natural elements remain – there are no complex organic residues.

Crop tolerance

All crop species are severely damaged or killed by direct application of ammonium sulphamate or by direct or indirect root-poisoning by percolation of the herbicide into the soil from treated stems and stumps. Cut stump application of ammonium sulphamate should therefore be restricted to pre-planting.

Product rate and methods of application

Apply to fresh cut stump surfaces.

Ammonium sulphamate may be applied in two ways:

a. A 40% solution of ammonium sulphamate (0.4 kg crystals per 1 litre of water) to the point of run-off, using a plastic watering can or a Tecnoma T18P semi-pressurised knapsack sprayer (see Section 11.5.4).

b. Dry ammonium sulphamate crystals at the rate of 6 g per 10 cm of stump diameter.

Timing of application

Best results are obtained from applications made during

June–September but ammonium sulphamate can be applied at any time between April and September.

Additional information

1. **Weed control**

a. Ammonium sulphamate is best applied in dry weather so that the spray solution or the crystals are not washed off the treated stump.

b. Solutions of ammonium sulphamate in water should be freshly prepared on each day of use.

2. **Protective clothing**

Read the product label for protective clothing and equipment requirements and check there are no items required in addition to the Forestry Commission recommendations in Section 10.

3. **Special precautions**

Ammonium sulphamate is harmful to fish; do not contaminate ponds, watercourses or ditches with the chemical or used container.

The label on the herbicide container has been designed for your protection – ALWAYS READ THE INSTRUCTIONS ON THE LABEL.

7.5.2 2,4-D/Dicamba/Triclopyr

Product

Broadshot 200:85:65 g/litre 2,4-D/dicamba/triclopyr (Cyanamid)

Description

A translocated herbicide which controls a wide range of annual and perennial herbaceous weeds, as well as woody weeds. Most grasses are resistant, although some may be suppressed.

Crop tolerance

Cut stump treatments may damage desirable trees and shrubs where their roots penetrate the treated area – only use as a pre-plant treatment.

Product rate

Apply a 15% solution of product in water.

Methods of application

Pre-plant

Apply to cover all exposed rim and collar bark on the freshly cut stump (heartwood need not be treated) by:

knapsack sprayer at low pressure;
forestry spot gun with a solid stream nozzle;
cleaning saw with suitable attachment.

To aid identification of treated stumps a suitable dye (e.g. red dye available from Hortichem Ltd.) may be added.

Refer to Section 11 for details of applicators.

Timing of application

At any time between felling and appearance of new growth.

Additional information

1. **Weed control**

a. Do not apply when rain is imminent, or in periods of very hot or cold weather.

b. Volatilisation can cause serious damage to neighbouring crops – if possible avoid applications in hot weather.

c. Any regrowth following application should be treated with a foliar spray of the product.

d. After cut stump treatments, a 3-month delay should be allowed before replanting within 1 metre of the treated area.

2. **Protective clothing**

   Read the product label for protective clothing and equipment requirements, and check there are no items required in addition to the Forestry Commission recommendations in Section 10.

3. **Special precautions**

a. Irritating to eyes and skin – avoid contact. Harmful if swallowed.

b. The herbicides can be dangerous to fish – do not contaminate ponds, watercourses or ditches with the chemical or used containers.

   The label on the herbicide container has been designed for your protection – ALWAYS READ THE INSTRUCTIONS ON THE LABEL.

7.5.3 Glyphosate

Products

Roundup Biactive	360 g/litre glyphosate (Monsanto)
Roundup Pro Biactive	360 g/litre glyphosate (Monsanto)

Description

A translocated herbicide normally applied to and taken up by the foliage but which is also effective when applied to freshly cut stumps. As a cut stump treatment is reduces or stops the production of new shoots and kills the root system of the stump.

Glyphosate controls all the major broadleaved woody weed species and is also highly effective in preventing regrowth from the remaining live branches on the stumps of SS and other conifers, which may be useful for respacing of natural regeneration.

Planting can be carried out 7 days after treatment.

Crop tolerance

There is little evidence of translocation across root grafts to untreated trees ('flashback'). In general, cut stumps can be safely treated by this method among any crop species, provided none of the herbicide is allowed to fall on the crop foliage, or on to exposed roots.

For foliar crop tolerance see Section 7.3.4.

Product rate

Conifer species and rhododendron: apply a 20% solution of the product in water.

Broadleaved species: apply a 10% solution of the product in water.

Methods of application

Apply to saturate the freshly cut stump by:

knapsack sprayer operated at low pressure;
forestry spot gun fitted with a solid stream nozzle;
cleaning saw fitted with a suitable herbicide spray attachment.

To aid identification of treated stumps a suitable dye (e.g. red dye, available from Hortichem Ltd.) may be added.

Refer to section 11 for details of applicators.

Timing of application

Best results are obtained when application is made to freshly cut stumps from October to February (outside the time of spring sap flow). Adequate control may also be possible from applications between late April to early August.

Application should be made within 1 week of felling.

Additional information

1. **Weed control**

a. Reduced weed control may result when glyphosate is applied to the surface of frozen stumps.

b. Diluted glyphosate may denature after 2 to 3 days if clean water is not used. Where possible use tap water as the diluent and only mix sufficient quantity for the day's programme.

2. **Protective clothing**

Read the product label for protective clothing and equipment requirements and check there are no items required in addition to the Forestry Commission recommendations in Section 10.

3. **Special precautions**

a. Unlike the additional glyphosate products listed at the end of this Section, Roundup Pro Biactive and Roundup Biactive DO NOT require a warning on the product label about harmful effects to aquatic life, or about irritant effects to operators. However, the addition of Mixture B makes any application potentially harmful to aquatic life, and a potential irritant to operators.

b. Do not contaminate ponds, watercourses or ditches with the chemical or used containers.

The label on the herbicide container has been designed for your protection – ALWAYS READ THE INSTRUCTIONS ON THE LABEL.

Additional glyphosate products with full forestry approval

Barclay Gallup	360 g/litre glyphosate (Barclay)
Barclay Gallup Amenity	360 g/litre glyphosate (Barclay)
Clayton Glyphosate	360 g/litre glyphosate (Clayton)
Clayton Swath	360 g/litre glyphosate (Clayton)
Glyphogan	360 g/litre glyphosate (PBI)
Glyphosate-360	360 g/litre glyphosate (Top Farm)
Helosate	360 g/litre glyphosate (Helm)

Hilite	144 g/litre glyphosate (Nomix-Chipman) – CDA formulation
Outlaw	360 g/litre glyphosate (Barclay)
Portman Glyphosate 360	360 g/litre glyphosate (Portman)
Roundup	360 g/litre glyphosate (Monsanto)
Roundup	360 g/litre glyphosate (Schering/Agro Evo)
Roundup Biactive Dry	42.6% w/w glyphosate (Monsanto)
Stacato	360 g/litre glyphosate (Unicrop)
Stefes Glyphosate	360 g/litre glyphosate (Stefes)
Stefes Kickdown 2	360 g/litre glyphosate (Stefes)
Stetson	360 g/litre glyphosate (Monsanto)
Stirrup	144 g/litre glyphosate (Nomix-Chipman) – CDA formulation

These products may have different conditions of use regarding operator and environmental safety – refer to the product label for further guidance.

7.5.4 Triclopyr

Products

Timbrel	480 g/litre triclopyr (DowElanco)
Garlon 4	480 g/litre triclopyr (DowElanco)
Chipman Garlon 4	480 g/litre triclopyr (Nomix-Chipman)

Description

A plant growth regulating herbicide which is rapidly absorbed, mainly through the foliage, but also by roots and stems. Once inside the plant it is readily translocated. Triclopyr is effective against most herbaceous and woody weeds.

Grasses sometimes show some yellowing following spraying operations, but this is quickly outgrown.

In the soil triclopyr is broken down fairly rapidly by microbial action.

Crop tolerance

There is no evidence of translocation across root grafts to untreated trees ('flashback'). Unwanted stems can be safely treated by this method among any crop species provided none of the herbicide is allowed to fall on the crop foliage.

For foliar crop tolerance see Section 7.3.6.

Product rate

Solution of product in water applied to the freshly cut stump.

For alder, birch, blackthorn, box and dogwood use a 2% solution (20 ml/l).
For hawthorn, laurel and rhododendron use an 8% solution.
For all other broadleaved tree species, use a 4% solution.

Methods of application

Apply to saturate the freshly cut surface of the stump and any remaining bark by:

knapsack sprayer operated at low pressure;
forestry spot gun fitted with a solid stream jet;
cleaning saw with suitable attachment.

So that the treated stumps can be identified during treatment a suitable dye (e.g. red dye available from Hortichem Ltd.) can be added to the spray solution.

Refer to Section 11 for details of applicators.

Timing of application

Best results are obtained when application is made to freshly cut stumps from October to March (outside the time of spring sap flow).

Applications should be made within 1 week of felling.

Additional information

1. **Weed control**
   Volatilisation can cause serious damage to neighbouring crops – if possible avoid applications in hot weather.
2. **Protective clothing**
   Read the product label for protective clothing and equipment requirements and check there are no items required in addition to the Forestry Commission recommendations in Section 10.
3. **Special precautions**

a. Triclopyr is harmful if in contact with skin and irritating to eyes – avoid contact.

b. Triclopyr is dangerous to fish; do not contaminate ponds, watercourses or ditches with the concentrate, spray drift or used containers.

The label on the herbicide container has been designed for your protection – ALWAYS READ THE INSTRUCTIONS ON THE LABEL.

8 Rhododendron

8.1 General

The glossy evergreen leaves of rhododendron and laurel have a thick waxy cuticle which is comparatively resistant to the entry of herbicides. To achieve adequate control application rates need to be higher than for most other woody weeds. Moreover, in older bushes translocation is very restricted and it is necessary to spray almost the whole of the leaf area and stem surfaces to achieve good control.

Rhododendron can be found on acid sites, mainly in the wetter western half of the country, in all phases of colonisation from a light scatter of small seedlings, through a partial cover of bushes, to impenetrable thickets 2–5 m in height.

The early stages of encroachment are (subject to terrain) easily accessible for herbicide application but bushes more than 1.5 m high must be manually or mechanically cleared to allow the stumps and the more susceptible regrowth to be sprayed, preferably before the regrowth is more than 1 m tall.

Herbicides approved for use in this situation are:
ammonium sulphamate
2,4-D/dicamba/triclopyr
glyphosate
imazapyr
triclopyr

These herbicides are applied either as a foliar spray or as a cut stump treatment, and frequently a combination of both techniques may be necessary where cut stumps and young regrowth exist together. All will give adequate control when spraying cut stumps or young regrowth, but when a foliar spray of larger bushes is the only option, imazapyr will give the best results.

All crops will be damaged by contact with these herbicides at the rates necessary to control rhododendron.

Section 7.5 gives details of cut-stump applications for all woody species – this section deals with foliar applications to rhododendron.

8.2 Ammonium sulphamate

Products

Amcide soluble crystals	100% ammonium sulphamate (Battle, Hayward and Bower)
Root-out soluble crystals	100% ammonium sulphamate (Dax)

Description

A highly soluble translocated, contact and soil-acting herbicide which is absorbed through leaves, roots and exposed live tissue surfaces. It is effective against most woody species including the more resistant species such as rhododendron, hawthorn and ash. It corrodes metals and alloy including copper, brass, mild steel and galvanised iron.

Break down in the soil can take up to 12 weeks during which time it retains its herbicidal properties. Three months should elapse between treatment and subsequent planting. After breakdown only natural elements remain – there are no complex organic residues.

Crop tolerance

All crop species are severely damaged or killed by direct application of ammonium sulphamate by absorption via the roots. Ammonium sulphamate should therefore only be used as a pre-planting treatment.

Product rates and methods of application

Apply a 40% solution of ammonium sulphamate (0.4 kg crystals per 1 litre of water) using a plastic watering can or a Tecnoma T18P semi-pressurised knapsack sprayer (see Section 11.5.4) to all accessible surfaces including freshly cut stumps and all remaining bark, twigs and foliage, wetting to the point of run-off.

Timing of application

Best results are obtained from applications made during the growing season with optimum control achieved with May and June treatments. However applications can be made anytime between April and September.

Additional information

1. **Weed control**

a. Ammonium sulphamate is best applied in dry weather so that the spray solution is not washed off treated surfaces.

b. Solution of ammonium sulphamate in water should be freshly prepared on each day of use.

c. Further applications may be necessary after 3–4 years to achieve a complete kill of rhododendron.

2. **Protective clothing**
   Read the product label for protective clothing and equipment requirements and check there are no items required in addition to the Forestry Commission recommendations in Section 10.

3. **Special precautions**
   Ammonium sulphamate is harmful to fish: do not contaminate ponds, watercourses or ditches with the chemical or used containers.

   The label on the herbicide container has been designed for your protection – ALWAYS READ THE INSTRUCTIONS ON THE LABEL.

8.3 2,4-D/Dicamba/Triclopyr

Product

Broadshot 200:85:65 g/litre 2,4-D/dicamba/triclopyr (Cyanamid)

Description

A translocated herbicide which controls a wide range of annual and perennial herbaceous weeds, as well as woody weeds. Most grasses are resistant, although some may be suppressed.

Crop tolerance

All crop species are likely to be damaged at the product rates required for rhododendron control. Ideally, pre-plant treatments should be used. Post-plant sprays may be possible if carefully directed away from the crop species foliage and rooting zone.

Product rate

Apply as a 7.5% solution of product in water.

Maximum product rate to prevent potential environmental pollution is 5 l/ha.

Methods of application

Pre-plant (overall) or post-plant (directed spray)

Tractor mounted equipment

Boom sprayer at LV or MV (150–400 1/ha).
Use the higher volume rate for larger, denser weed populations.

Handheld equipment

Knapsack sprayer at LV or MV.
Forestry spot gun at LV.

Refer to Section 11 for details of applicators.

Timing of application

Apply in July–September, after spring growth has slowed but before senescence.

Additional information

1. Weed control

a. Do not apply when rain is imminent, or in periods of very hot or cold weather.

b. Volatilisation can cause serious damage to neighbouring crops – if possible avoid applications in hot weather.

c. After spraying woody vegetation, 3 months delay should be allowed before replanting.

2. **Protective clothing**
   Read the product label for protective clothing and equipment requirements and check there are no items required in addition to the Forestry Commission recommendations in Section 10.

3. **Special precautions**

a. Irritating to eyes and skin – avoid contact. Harmful if swallowed.

b. The herbicide can be dangerous to fish – do not contaminate ponds, watercourses or ditches with the chemical or used containers.

   The label on the herbicide container has been designed for your protection – ALWAYS READ THE INSTRUCTIONS ON THE LABEL.

8.4 Glyphosate

Products

Roundup Biactive	360 g/litre glyphosate (Monsanto)
Roundup Pro Biactive	360 g/litre glyphosate (Monsanto)

Description

A translocated herbicide taken up by the foliage and conveyed to the roots. It causes chlorosis and eventual death of leaves and kills roots and shoots.

With most species, once inside the plant it is readily translocated, but translocation within rhododendron seems to be particularly poor in a tangential direction, so that spraying part of a bush results in the death of that part only.

Glyphosate controls a wide range of weeds including grasses, herbaceous broadleaved weeds, bracken, heather and woody weeds. With the latter group there may be little effect until the following season when roots are killed and re-suckering prevented.

On contact with the soil glyphosate is quickly inactivated. Planting can be carried out 7 days after treatment, and a minimum of 7 days should be allowed before cultivation and the breaking up of rhizomes and roots.

Crop tolerance

At the rates of application needed to kill rhododendron, all forestry crop species are severely damaged by overall treatments of glyphosate. Ideally all rhododendron control should be carried out pre-planting but, if a crop is present, treatment is possible if the spray is very carefully directed to avoid contact with crop trees.

Product rate

Apply 10 litres of product per hectare; *or* 8 litres of product per hectare plus Mixture B at 2% of final spray volume.

Alternatively spray so that all the foliage is wetted, but the herbicide solution does not run off, with a 2% (2 litres of product in 100 litres water) solution using a knapsack sprayer. For cut stump treatments follow the recommendations in Section 7.5.3.

Methods of application

Pre-plant (overall application) or post-plant (directed)

Tractor mounted equipment

Boom sprayer at LV or MV.

Ulvaforest low speed rotary atomiser at VLV.

Handheld equipment

Knapsack sprayer at MV.

Knapsack sprayer with 'VLV' nozzle at LV.

Herbi low speed rotary atomiser or ULVA high speed rotary atomiser at VLV: for full effect dilute the herbicide with at least 3 times the volume of water.

Forestry spot gun at LV (only for seedling rhododendron).

Refer to Section 11 for details of applicators and correct calibration.

Timing of application

June–September.

Additional information

1. **Weed control**

a. Glyphosate applied later than June will be too late to lessen the effect of weed competition in the current season.

b. Glyphosate is most effective when relative humidity is high and the air is warm (e.g. 15°C+).

c. Reduced weed control may result when plants are under stress, e.g. drought.

d. Heavy rainfall within 24 hours of application may reduce the herbicide's effectiveness by preventing sufficient foliar absorption. The addition of Mixture B at 2% of final spray volume will improve reliability in these circumstances.

e. Diluted glyphosate may denature after 2 to 3 days if clean water is not used. Where possible use tap water as the diluent and only mix sufficient for the day's programme.

2. **Protective clothing**

Read the product label for protective clothing and equipment requirements and check there are no items required in addition to the Forestry Commission recommendations in Section 10.

3. **Special precautions**

a. Unlike the additional glyphosate products listed at the end of this Section, Roundup Pro Biactive and Roundup Biactive DO NOT require a warning on the product label about harmful effects to aquatic life, or about irritant effects to operators. However, the addition of Mixture B makes any application potentially harmful to aquatic life, and a potential irritant to operators.

b. Do not contaminate ponds, watercourses or ditches with the chemical or used containers.

Additional glyphosate products with full forestry approval

Product	Formulation
Barclay Gallup	360 g/litre glyphosate (Barclay)
Barclay Gallup Amenity	360 g/litre glyphosate (Barclay)
Clayton Glyphosate	360 g/litre glyphosate (Clayton)
Clayton Swath	360 g/litre glyphosate (Clayton)
Glyphogan	144 g/litre glyphosate (PBI)
Glyphosate-360	360 g/litre glyphosate (Top Farm)
Helosate	360 g/litre glyphosate (Helm)
Hilite	144 g/litre glyphosate (Nomix-Chipman) – CDA formulation
Outlaw	360 g/litre glyphosate (Barclay)
Portman Glyphosate 360	360 g/litre glyphosate (Portman)
Roundup	360 g/litre glyphosate (Monsanto)
Roundup	360 g/litre glyphosate (Schering/Agro Evo)
Roundup Biactive Dry	42.6% w/w glyphosate (Monsanto)
Stacato	360 g/litre glyphosate (Unicrop)
Stefes Glyphosate	360 g/litre glyphosate (Monsanto)
Stefes Kickdown 2	360 g/litre glyphosate (Stefes)
Stetson	360 g/litre glyphosate (Monsanto)
Stirrup	144 g/litre glyphosate (Nomix-Chipman) – CDA formulation

These products may have different conditions of use regarding operator and environmental safety – refer to the product label for further guidance.

8.5 Imazapyr

Product

Arsenal 50F 50 g/litre imazapyr (Cyanamid, marketed by Nomix-Chipman)

Description

A translocated and residual herbicide which will control a wide range of established annual and perennial grass and herbaceous broadleaved weeds, and will provide long term residual control of germinating seedlings. Control of woody weeds is likely to be good, although there are no manufacturer's label recommendations at present. Rhododendron control is good, and a measure of long term control is given. Imazapyr is absorbed through roots and foliage, then rapidly translocated throughout the plant. Treated plants stop growing soon after application, but chlorosis and tissue necrosis may not be apparent until up to 4–6 weeks after application – complete kill may take several weeks.

Crop tolerance

Imazapyr should only be used as a pre-plant treatment. The only tolerant species according to the product label are SS, LP and CP. Limited FC trials suggest JL, DF, and NS may also be tolerant. The trees must be at least 2 year old nursery stock, and 5 months must elapse between treatment and subsequent planting. All other species may be damaged.

Product rate

Apply overall to all foliage at 4 litres of product per treated hectare.

Apply at 15 litres of product per treated hectare where only partial treatment of larger bushes is possible.

Use Mixture B at a 2% solution of final spray volumes for all applications. Choose a volume rate to achieve the point of run-off and calibrate accordingly.

Refer to Section 11 for details of applicator and correct calibration.

Methods of application

Pre-plant (overall)

Tractor mounted equipment

Boom sprayer at LV-MV (200 l/ha).
Ulvaforest low speed rotary atomiser at VLV.

Handheld equipment

Knapsack sprayer at LV-MV (200 l/ha).

Forestry spot gun at LV.

Timing of application

For best results apply outside the winter season, from April–October.

Additional information

1. **Weed control**

a. Most effective control will be gained from overall applications to all foliage, but larger bushes can be killed if at least ¼ of the foliage is treated at the higher application rate, although long-term control is reduced.

   There is no information at present on cut-stump treatments, although control is likely to be good.

b. Do not use imazapyr on soils that may later be used for growing desirable plants, except for the conifers species SS, CP, and LP.

c. Do not use on conifer nursery beds.

d. Do not use near desirable trees or shrubs, nor in areas into which their roots may extend or in locations where the chemical may be washed or moved into contact with their roots.

e. Do not apply in windy conditions, or using high pressures producing a fine spray liable to drift.

f. The length of residual weed control given has not been fully evaluated, but indications are that in some circumstances weed suppression may still occur up to two seasons after planting. However, this may vary from site to site, and there are no manufacturer's recommendations at present.

g. Leave at least 48 hours from treating bushes, before any cutting operations.

h. Results of trials suggest that planted trees are unlikely to be damaged from root contact with treated vegetation, or from leachates from dead vegetation that is subsequently cultivated. However, imazapyr may be very mobile in soil.

2. **Protective clothing**

   Read the product label for protective clothing and equipment requirements and check there are no items required in addition to the Forestry Commission recommendations in Section 10.

3. Special precautions

a. Irritating to the eyes – avoid contact.

b. Do not contaminate ponds, watercourses or ditches with the chemical or used containers.

The label on the herbicide container has been designed for your protection – ALWAYS READ THE INSTRUCTIONS ON THE LABEL.

8.6 Triclopyr

Products

Timbrel 480 g/litre triclopyr (DowElanco)

Garlon 4 480 g/litre triclopyr (DowElanco)

Chipman Garlon 4 480 g/litre triclopyr (Nomix-Chipman)

Description

A plant growth regulating herbicide which is rapidly absorbed, mainly through the foliage, but also by roots and stems. With most species, once inside the plant it is readily translocated, but translocation within Rhododendron seems to be particularly poor in a tangential direction, so that spraying part of a bush results in the death of that part only. Most herbaceous and woody species are controlled.

Grasses sometimes show some yellowing following spraying operations but this is quickly outgrown.

In the soil triclopyr is broken down fairly rapidly by microbial action.

Planting should be deferred for at least 6 weeks after application.

Crop tolerance

At the rates of application needed to kill rhododendron all crop species are likely to be damaged by overall treatments of triclopyr. Ideally all rhododendron control should be carried out pre-planting but, if a crop is present, treatment is possible if the spray is very carefully directed to avoid contact with crop trees.

Product rate

Apply 8 litres of product per treated hectare in water.

Alternatively, for taller denser rhododendron, spray so that all the foliage is wetted but the herbicide does not run off, using a mixture of 130 ml Timbrel in 5 litres of water (130 ml Timbrel + 4870 ml water.

Methods of application

Pre-plant (overall) or post-plant (directed)

Tractor mounted equipment

Tractor mounted boom sprayer at MV.

Handheld equipment

Knapsack sprayer MV.

Refer to Section 11 for details of applicators and correct calibration.

Timing of application

June–September.

Additional information

1. **Weed control**

a. Volatilisation can cause serious damage to neighbouring crops – if possible avoid applications in hot weather.

b. Rainfall within 2 hours of application may reduce the herbicide's effectiveness by preventing sufficient foliar absorption.

c. A minimum interval of 6 weeks is required between applications of triclopyr and planting.

2. **Protective clothing**

Read the product label for protective clothing and equipment requirements and check there are no items required in addition to the Forestry Commission recommendations in Section 10.

3. **Special precautions**

a. Triclopyr is harmful if in contact with skin and irritating to eyes – avoid contact.

b. Triclopyr should not be applied via controlled droplet applicators.

c. Triclopyr is dangerous to fish: do not contaminate ponds, watercourses or ditches with the concentrate, spray drift or used containers.

The label on the herbicide container has been designed for your protection – ALWAYS READ THE INSTRUCTIONS ON THE LABEL.

9 Farm forestry weed control only

9.1 General

The herbicides in this section all have specific off-label or label approval for use on land previously under arable cultivation, or improved grassland that is being converted to farm woodland or short rotation coppice, (applied for by the Forestry Commission). Where these uses are not specified on the label, applications are at the user's own risk. This means that the manufacturers cannot be held responsible for any adverse effects on crops or failure to control weeds, while employers and operators still have the responsibilities that apply when using the products' on-label recommendations regarding dose rate, spray volume, etc.

OFF-LABEL OR LABEL APPROVAL FOR USE IN FARM FORESTRY ESTABLISHMENT DOES NOT IMPLY APPROVAL FOR USE IN CONVENTIONAL FORESTRY SITUATIONS.

Where one exists, the relevant off-label approval for each product is reprinted after the individual herbicide section. However, all the products with forestry approval listed in the previous sections of the book can be used in farm forestry situations. The products listed in this section may be of use where vigorous re-growth of arable weeds on fertile ex-agricultural sites is anticipated. In Forestry Commission trials, some of these herbicides were found to be safe for overspraying some tree species – see Table 7. However, it should be noted that these data were obtained in small-scale trials – there is always a risk of damage when overspraying actively growing trees. All applications are at user's own risk. Users are advised to conduct limited trials of new herbicides before adopting them on a commercial scale.

Under the new long-term off-label arrangements, products with approval for use in cereal crops may be used in farm woodland situations for the first 5 years after planting (see Section 2.4). This arrangement does not allow the use of any of the additional products that the Forestry Commission are testing and for which specific off-label approval is awaited. Only products which have full label approval, or have been subject to test prior to specific off-label applications, are included in this Section. However, users should be aware that further products with which they are familiar may become available for use in certain farm woodland situations, under the long-term off-label arrangements.

A revised version of Forestry Commission Research Information Note 201 will detail some of the additional herbicides that may be of use in short rotation coppice situations, due to the long-term off-label arrangements.

No detailed information on appropriate applicators is given in this section – it is assumed that farmers will have access to a range of applicators outside the scope of this Field Book (refer to Technical Development Branch Note 8/94 for an introduction to the subject), although the range of applicators listed in Section 11 are still appropriate.

The most appropriate herbicide for farm forestry situations can be identified by referring to individual herbicide sections, crop tolerance (Table 7) and weed susceptibility (Table 8). Tank mixes of chemicals are often used in agriculture – information on this is given in Table 9.

Information on clopyralid, isoxaben and propyzamide is included in the tables in this Section, even though they are covered by a forestry off-label approval in the first case, and full forestry approvals for the others, because of their potential usefulness when combined with other products in farm woodland situations. Refer to Section 4 for more detail on these three herbicides.

Pre-emergent soil acting herbicides should be applied after trees are planted, prior to bud burst and weed emergence. In subsequent years they should be applied to bare soil in early spring (except for propyzamide), before weed emergence. If correctly applied these products should give adequate weed control for one season, but metazachlor and cyanazine are less persistent that propyzamide, pendimethalin and isoxaben, having an effective life of about 12 weeks. Repeat applications of metazachlor and isoxaben are possible, but weeds should only be treated if they are still at a susceptible stage of growth. Rain is necessary after applications to move the herbicides into the top 2–3 cm of the soil. Applications should be made to a firm, fine tilth – if large clods are present at the time of application these will weather and crumble, exposing untreated soil and allowing prolific weed growth.

Table 7 Crop tolerance to farm forestry herbicides

Herbicide	Dormant		Post-flushing[5] Trees will be at their most sensitive immediately after flushing. Herbicide application should not be made before new needles/ leaves have hardened.	
	Conifers	Broadleaves	Conifers	Broadleaves
metazachlor	✓	✓	✓[3]	✓
clopyralid	✓	✓	✓	✓
isoxaben	✓	✓	✓	✓
cyanazine	✓	✓	✓	X
fluazifop-p-butyl	✓	✓	✓	✓
propyzamide	✓	✓	✓[4]	✓[4]
pendimethalin	✓	✓	✓	✓
propaquizafop[6]	✓	✓	✓	✓

✓ = herbicides can be used as an overall spray
X = herbicides must be applied as a directed spray

Notes:

1. The Table summarises results of FC trials – when used as listed above the herbicides can be used to treat the species listed on the next page.
2. For the purpose of this Table, treat larch as a broadleaved tree.
3. In Forestry Commission experiments where metazachlor was applied to pine in active growth (e.g. candles fully extended but needles not fully hardened) damage was seen. The damage symptoms were distortion, browning and loss of needles from the tender new growth. On a few plants the growing tip was killed, but on most plants the main stem (or candle) remained healthy but devoid of needles. Terminal buds were set as normal.
4. By the time trees have flushed it is too late to achieve weed control using propyzamide.
5. In Forestry Commission trials, the treatments listed as safe to overspray trees posting-flushing were found to have no significant effect on height or survival of the crop species listed. However, there may be some transient foliage damage. Where condition of foliage is particularly important, such as in Christmas trees, overall post-flushing applications are not recommended.

Conifers	Broadleaves
Sitka spruce	Oak
Norway spruce	Ash
Douglas fir	Sycamore
Noble fir	Beech
Corsican pine	Wild cherry
Western red cedar	Birch
Japanese larch[2]	Alder
Scots pine	Sweet chestnut
	Norway maple
	Poplar (sets)
	Willow (sets)

6. Propaquizafop was not subject to Forestry Commission trials – data are based on manufacturer's trials. Propaquizafop has NOT been tested on Norway maple.

Table 8 Susceptibility of common arable weeds to selective farm forestry herbicides

Weeds	Pre-emergent herbicide					Post-emergent herbicide					
	metazachlor	isoxaben	cyanazine	propyzamide	pendimethalin	metazachlor	cyanazine	clopyralid	fluazifop-p-butyl	propaquizafop	propyzamide†
Bents				S					4ELT		S
Bitter cress, hairy		S									
Bittersweet				S							MS
Black bindweed	MS		S	S	S		100mm	2ETL			MS
Black grass	S		MS	S	S	2ETL	2ETL		FT	FT	S
Brome, barren	MS		S	S			2ETL		FT		S
Buttercup, corn		S			S						
Buttercup, creeping				S							MS
Canary grass, awned					S						
Chamomile, corn	S	S			S						
Chamomile, stinking	S	S			S			2ETL			
Charlock	MR	S	S				6ETL				
Chickweed, common	S	S	S	S	S	4ETL	100mm				S
Cleavers	MS	MS	MR	S	S*						R
Clover (from seed)								2ETL			
Coltsfoot				S							MR
Common couch				S					4ETL	3ETL	S
Crane's-bill, cut-leaved	S					C					
Creeping bent (watergrass)				S					4ETL		S
Creeping soft grass		R		S							MS

Table 8 Susceptibility of common arable weeds to selective farm forestry herbicides – *continued*

Weeds	Pre-emergent herbicide					Post-emergent herbicide					
	metazachlor	isoxaben	cyanazine	propyzamide	pendimethalin	metazachlor	cyanazine	clopyralid	fluazifop-p-butyl	propaquizafop	propyzamide†
Crested dog's tail				S							S
Curled dock				S							MS
Dead-nettle, henbit			S		S		100mm				
Dead-nettle, red	S	S	S		S	2ETL	100mm				
Dead-nettle, white			S		S		100mm				
Dock, broadleaved				S							MS
Established perennials			R								
False oat grass				S							S
Fat-hen	MS	S	MS	S	S		2ETL				MS
Fescue, meadow				S							S
Fescues				S							S
Field horsetail				S							MS
Fleabane, common	MS										
Fool's parsley			MR				1ETL				
Forget-me-not, field	S	S	S	S	S	2ETL	4ETL				S
Foxglove				R							R
Fumitory, common	R	S	MS	MS	S		1ETL				R
Gromwell, field	S										
Groundsel	S	S	S			2ETL	1ETL	6ETL			
Hemp-nettle, common	MR	S	S		S		100mm				
Knotgrass	R	S	MS	S	S		1ETL				MS

Table 8 Susceptibility of common arable weeds to selective farm forestry herbicides – *continued*

Weeds	Pre-emergent herbicide					Post-emergent herbicide					
	metazachlor	isoxaben	cyanazine	propyzamide	pendimethalin	metazachlor	cyanazine	clopyralid	fluazifop-p-butyl	propaquizafop	propyzamide†
Marigold, common	S	S			S	2ETL		6ETL			
Mat grass				S							S
Mayweed, scented	S	S	S		S	4ETL	2ETL	6ETL			
Mayweed, scentless	S	S	S		S	4ETL	2ETL	6ETL			
Meadow foxtail				S							S
Meadow grass, annual	S	R	S	S	S	2ETL	FT			3ETL	S
Meadow grass, rough		R	S	S	S		FT				S
Meadow grass, smooth		R		S							S
Mustard, black	MR		S				6ETL				
Mustard, white	MR		S				6ETL				
Nettle, small	MS	S	S	S	S		100mm				MS
Nightshade, black			MS	S	S		100mm				MS
Onion, couch									4ETL		
Orache, common	S	S	S		S		1ETL				
Pale persicaria			S				2ETL				
Pansy, field	MR	S	MS		S		1ETL				
Pansy, wild	MR				S						
Parsley piert	S	S	S		S		1ETL				
Pimpernel, scarlet		S	S		S		100mm				
Pineapple weed	S	S			S	4ETL		6ETL			
Poppy, common	S	S	S		S						

Table 8 Susceptibility of common arable weeds to selective farm forestry herbicides – *continued*

Weeds	Pre-emergent herbicide					Post-emergent herbicide					
	metazachlor	isoxaben	cyanazine	propyzamide	pendimethalin	metazachlor	cyanazine	clopyralid	fluazifop-p-butyl	propaquizafop	propyzamide†
Purple moor grass				S							S
Radish, wild		S	S				100mm				
Redshank	MS	S	S	S	S		100mm				MS
Rosebay willowherb				R							R
Rye grasses			S	S			3ETL		FT	FT	S
Sedges				MS							MS
Sheep's sorrel				S							MS
Shepherd's purse	S	S	S	MS	S		100mm				R
Soft brome				S							S
Speedwell, common	S	S	S	S	S	2ETL	100mm				MS
Speedwell, germander	S	S	S		S	2ETL	100mm				
Speedwell, green	S	S	S		S	2ETL	100mm				
Speedwell, grey	S	S	S		S	2ETL	100mm				
Speedwell, ivy-leaved	S	S	S	S	S	2ETL	100mm				MS
Speedwell, wall	S	S	S		S	2ETL	100mm				
Spurrey, corn	MS	S		S							S
Sweet vernal grass				S							S
Thale cress		S									
Thistle, creeping								250mm			
Thistle, smooth sow (perennial)						S		6ETL			
Thistle, spear								250mm			

9

Table 8 Susceptibility of common arable weeds to selective farm forestry herbicides – *continued*

Weeds	Pre-emergent herbicide					Post-emergent herbicide					
	metazachlor	isoxaben	cyanazine	propyzamide	pendimethalin	metazachlor	cyanazine	clopyralid	fluazifop-p-butyl	propaquizafop	propyzamide†
Timothy				S							S
Trefoils (from seed)								2ETL			
Tufted hair grass				S							S
Vetches (from seed)								2ETL			
Volunteer cereals	MR			S					FT	FT	S
Volunteer oilseed rape	R	S			S*						
Wavy hair grass				S							S
Wild carrot								2ETL			
Wild oat	MR			S	S				FT	FT	S
Wood small reed				S							S
Yellow oat grass				S							S
Yorkshire fog				S							S

Key: Pre-emergent

S – susceptible
MS – moderately susceptible
MR – moderately resistant
R – resistant
– not tested
* – plant arising from deep-germinating seeds may not be controlled

Key: Post-emergent
Growth stage of weeds (latest at which controlled):

C – cotyledon stage
ETL – number of expanded true leaves
ETLs – number of expanded true leaves (suppression only)
mm – diameter or height of weeds
Fbv – flower bud visible
FT – fully tillered

† – all weed susceptibilities for propyzamide post emergence are for fully established weeds

Note: Table 8 is reproduced by kind permission of the herbicide manufacturers

Herbicide mixtures

1. In agriculture, control of the wide range of weeds found on arable sites is commonly achieved using tank mixtures of herbicides. The application of tank mixes may be appropriate during tree establishment to cope with difficult mixtures of weeds on similar site types.
2. Results of trials carried out during 1989 and 1990 at a number of sites in southern Britain indicate that the herbicide mixtures listed in Table 9 are tolerated by the coniferous and broadleaved species listed in Table 7 when applied as overall sprays before bud burst in the spring. Propaquizafop was not subject to Forestry Commission trials – entries are based on the manufacturer's data.

Table 9 Farm forestry herbicide tank mixes (all herbicides at approved rate)

Propyzamide	metazachlor isoxaben cyanazine clopyralid pendimethalin
Metazachlor	clopyralid isoxaben propyzamide pendimethalin
Cyanazine	clopyralid isoxaben pendimethalin
Isoxaben	metazachlor clopyralid cyanazine propyzamide pendimethalin
Propaquizafop	metazachlor clopyralid

3. Mixtures of herbicides in the left column, with those listed to their right, were found to be safe, and are allowed under current regulations – see Section 3.7. Users should note however that any mixtures not specified on the relevant product label are undertaken at their own risk.

9

9.2 Cyanazine

Product

Fortrol 500 g/litre cyanazine (Cyanamid)

Description

A soil and foliar acting herbicide for the control of annual dicotyledons and annual grasses in farm forestry. See Table 8.

Crop tolerance

See Table 7.

Product rate

Apply 4.0 litres of product per treated hectare.

Methods of application

Pre- or post-plant (overall, band, spot or directed)

Knapsack or tractor mounted sprayers at MV or LV.

Timing of application

Best results will be achieved from applications after planting, before bud burst and weed germination. Control may last up to 12 weeks. One application a year is allowed.

Additional information

1. Weed control

a. For best results apply to a brash and weed free soil, cultivated to achieve a firm, fine, moist tilth.

b. Rain is required after application for effective weed control.

c. The extent of residual control will be reduced by soils with greater than 10% organic matter content (i.e. peat, brash or litter content), and with high soil temperatures.

d. Avoid applications in dull, cold or wet conditions.

e. Do not use on trees grown for Christmas tree production – some damage to needles is possible.

f. A tank mix of 4 litres per hectare of cyanazine with 5 litres per hectare of atrazine (see Section 4.2) will give the equivalent rates of active ingredient per hectare as the old Holtox product. This will give control of a wider range of grass species than cyanazine alone.

2. Protective clothing

Read the product label for protective clothing and equipment requirements and check there are no items required in addition to the Forestry Commission recommendations in Section 10.

3. Special precautions

a. Harmful if swallowed or in contact with the skin.

b. Harmful to fish – do not contaminate ponds, watercourses or ditches with the chemical or the used container.

The label on there herbicide container has been designed for your protection – ALWAYS READ THE INSTRUCTIONS ON THE LABEL.

NOTICE OF APPROVAL No. 0603/94

FOOD AND ENVIRONMENT PROTECTION ACT 1985
CONTROL OF PESTICIDES REGULATIONS 1986
(S.I. 1986 No. 1510):
APPROVAL FOR OFF-LABEL USE OF AN APPROVED PESTICIDE PRODUCT

This approval provides for the use of the product named below in respect of crops and situations, other than those included on the product label. Such 'off-label use' as it is known is at all times done at the user's choosing, and the commercial risk is entirely his or hers.

The conditions below are statutory. They must be complied with when the off-label use occurs. Failure to abide by the conditions of approval may constitute a breach of that approval, and a contravention of the Control of Pesticides Regulations 1986. The conditions shown below supersede any on the label *which would otherwise apply*.

Level and scope:	In exercise of the powers conferred by regulation 5 of the Control of Pesticides Regulations 1986 (SI 1986/1510) and of all other powers enabling them in that behalf, the Minister of Agriculture, Fisheries and Food and the Secretary of State, hereby jointly give full approval for the use of
Product name:	Fortrol containing
Active ingredient:	500 g/l cyanazine
Marketed by:	Shell Chemicals UK Ltd under MAFF No. 00924 subject to the conditions relating to off-label use set out below:
Date of issue:	6 April 1994
Date of expiry:	Unlimited (subject to the continuing approval of MAFF 00924)

Field of use:	ONLY AS A FORESTRY HERBICIDE
Situations:	Farm forestry
Maximum individual dose:	4 litres product/hectare
Maximum number of treatments:	One per year
Operator protection:	(1) Engineering control of operator exposure must be used where reasonably practicable in addition to the following personal protective equipment: Operators must wear suitable protective clothing (coveralls), suitable protective gloves and face protection (faceshield) when handling the concentrate. (2) However, engineering controls may replace personal protective equipment if a COSHH assessment shows they provide an equal or higher standard of protection.
Environmental protection:	Since this product is harmful to fish or aquatic life, surface waters or ditches must not be contaminated with chemical or used container.
Other specific restrictions:	(1) This product must only be applied if the terms of this approval, the product label and/or leaflet and any additional guidance on off-label approvals have first been read and understood.

9

(2) Crops grown within the treated areas of woodland must not be used for human or animal consumption.

(3) This product must only be used for forestry establishment on land previously under arable cultivation or improved grassland.

Signed J Micklewright
(Authorised signatory)

Date 6 April 1994

Application Reference Number: COP 93/01119

THIS NOTICE OF APPROVAL IS NUMBER 0603 of 1994

THIS NOTICE OF APPROVAL SUPERSEDES NOTICE OF APPROVAL No. 0192 OF 1991

ADVISORY INFORMATION

This approval relates to the use of 'Fortrol' in areas of conversion from arable land or improved grassland to farm forestry and coppice woodland, to be applied via conventional ground-based machinery (vehicle mounted hydraulic sprayers and knapsack sprayers).

Please note that an identical approval has been issued for this off-label use for 'Fortrol', marketed by Cyanamid UK Ltd, MAFF 07009.

NOTICE OF APPROVAL No. 0602/94

FOOD AND ENVIRONMENT PROTECTION ACT 1985
CONTROL OF PESTICIDES REGULATIONS 1986
(S.I. 1986 No. 1510):
APPROVAL FOR OFF-LABEL USE OF AN APPROVED PESTICIDE PRODUCT

This approval provides for the use of the product named below in respect of crops and situations, other than those included on the product label. Such 'off-label use' as it is known is at all times done at the user's choosing, and the commercial risk is entirely his or hers.

The conditions below are statutory. They must be complied with when the off-label use occurs. Failure to abide by the conditions of approval may constitute a breach of that approval, and a contravention of the Control of Pesticides Regulations 1986. The conditions shown below supersede any on the label *which would otherwise apply.*

Level and scope:	In exercise of the powers conferred by regulation 5 of the Control of Pesticides Regulations 1986 (SI 1986/1510) and of all other powers enabling them in that behalf, the Minister of Agriculture, Fisheries and Food and the Secretary of State, hereby jointly give full approval for the use of
Product name:	Fortrol containing
Active ingredient:	500 g/l cyanazine
Marketed by:	Cyanamid UK Limited under MAFF No. 07009 subject to the conditions relating to off-label use set out below:
Date of issue:	6 April 1994
Date of expiry:	Unlimited (subject to the continuing approval of MAFF 07009)

Field of use:	ONLY AS A FORESTRY HERBICIDE
Situations:	Farm forestry
Maximum individual dose:	4 litres product/hectare
Maximum number of treatments:	One per year
Operator protection:	(1) Engineering control of operator exposure must be used where reasonably practicable in addition to the following personal protective equipment: Operators must wear suitable protective clothing (coveralls), suitable protective gloves and face protection (faceshield) when handling the concentrate. (2) However, engineering controls may replace personal protective equipment if a COSHH assessment shows they provide an equal or higher standard of protection.
Environmental protection:	Since this product is harmful to fish or aquatic life, surface waters or ditches must not be contaminated with chemical or used container.
Other specific restrictions:	(1) This product must only be applied if the terms of this approval, the product label and/or leaflet and any additional guidance on off-label approvals have first been read and understood. (2) Crops grown within the treated area of woodland

must not be used for human or animal consumption.

(3) This product must only be used for forestry establishment on land previously under arable cultivation or improved grassland.

Signed J Micklewright
(Authorised signatory)

Date 6 April 1994

Application Reference Number: COP 93/01119

THIS NOTICE OF APPROVAL IS NUMBER 0602 of 1994
THIS NOTICE OF APPROVAL SUPERSEDES NOTICE OF APPROVAL No. 0192 OF 1991

ADVISORY INFORMATION

This approval relates to the use of 'Fortrol' in areas of conversion from arable land or improved grassland to farm forestry and coppice woodland, to be applied via conventional ground-based machinery (vehicle mounted hydraulic sprayers and knapsack sprayers).

Please note that an identical approval has been issued for this off-label use for 'Fortrol', marketed by Shell Chemicals UK Ltd, MAFF 00924.

9.3 Fluazifop-p-butyl

Products

Fusilade 5	125 g/litre fluazifop-p-butyl (Zeneca)
Fusilade 250 EW	250 g/litre fluazifop-p-butyl (Zeneca)

Description

A foliar acting herbicide for the control of annual and perennial grasses in farm forestry. This herbicide will have little effect on herbaceous broadleaved weeds. See Table 8.

Note that at the time of writing, Fusilade 5 is being phased out in favour of Fusilade 250 EW, A DOUBLE STRENGTH FORMULATION. Off-label approval for this new product has been obtained.

Crop tolerance

See Table 7.

Product rate

Apply 3.0 litres of product per treated hectare for Fusilade 5.

Apply 1.5 litres of product per treated hectare for Fusilade 250EW.

Methods of application

Pre- or post-plant (overall, band or directed)

Apply ONLY through tractor mounted sprayers, at MV. Do not apply through handheld applicators.

Timing of application

Timing of application is dependent on the growth stage of the target weed, and the crop present – see Tables 7 and 8.

Two applications per year can be made.

Additional information

1. **Weed control**

a. Always apply with agral wetter at the rate of 1 litre per 1000 litres of diluted spray.

b. Best results will be achieved by applications when weeds are actively growing under warm conditions with adequate soil moisture.

2. **Protective clothing**

Read the product label for protective clothing and equipment requirements and check there are no items required in addition to the Forestry Commission recommendations in Section 10.

3. Special precautions

a. Irritating to eyes and skin – avoid contact.

b. Harmful to fish – do not contaminate ponds, watercourses or ditches with the chemical or the used container.

The label on there herbicide container has been designed for your protection – ALWAYS READ THE INSTRUCTIONS ON THE LABEL.

NOTICE OF APPROVAL No. 0452/94

FOOD AND ENVIRONMENT PROTECTION ACT 1985
CONTROL OF PESTICIDES REGULATIONS 1986
(S.I. 1986 No. 1510):
APPROVAL FOR OFF-LABEL USE OF AN APPROVED PESTICIDE PRODUCT

This approval provides for the use of the product named below in respect of crops and situations, other than those included on the product label. Such 'off-label use' as it is known is at all times done at the user's choosing, and the commercial risk is entirely his or hers.

The conditions below are statutory. They must be complied with when the off-label use occurs. Failure to abide by the conditions of approval may constitute a breach of that approval, and a contravention of the Control of Pesticides Regulations 1986. The conditions shown below supersede any on the label *which would otherwise apply.*

Level and scope:	In exercise of the powers conferred by regulation 5 of the Control of Pesticides Regulations 1986 (SI 1986/1510) and of all other powers enabling them in that behalf, the Minister of Agriculture, Fisheries and Food and the Secretary of State, hereby jointly give full approval for the use of
Product name:	Fusilade 5 containing
Active ingredient:	125 g/l fluazifop-p-butyl
Marketed by:	Zeneca Ltd under MAFF No. 06669 subject to the conditions relating to off-label use set out below:
Date of issue:	7 March 1994
Date of expiry:	Unlimited (subject to the continuing approval of MAFF 06669)

Field of use:	ONLY AS A AGRICULTURAL/ HORTICULTURAL/FORESTRY HERBICIDE
Situations:	Farm forestry
Maximum individual dose:	3 litres product/hectare
Maximum number of treatments:	Two per year
Operator protection:	(1) Engineering control of operator exposure must be used where reasonably practicable in addition to the following personal protective equipment: (a) Operators must wear suitable protective clothing (coveralls), suitable protective gloves and face protection (faceshield) when handling the concentrate. (b) Operators must wear suitable protective clothing (coveralls), suitable protective gloves when spraying and adjusting or maintaining equipment or handling contaminated surfaces. (2) However, engineering controls may replace personal protective equipment if a COSHH assessment shows they provide an equal or higher standard of protection.
Environmental protection:	Since this product is dangerous to fish or aquatic life, surface

9

waters or ditches must not be contaminated with chemical or used container.

Other specific restrictions:

(1) This product must only be applied if the terms of this approval, the product label and/or leaflet and any additional guidance on off-label approvals have first been read and understood.

(2) This product must only be applied by handheld equipment.

(3) This product must only be used for forestry establishment on land previously under arable cultivation or improved grassland.

Signed J Micklewright
(Authorised signatory)

Date 7 March 1994

Application Reference Number: COP 93/01116

THIS NOTICE OF APPROVAL IS NUMBER 0452 of 1994
THIS NOTICE OF APPROVAL SUPERSEDES NOTICE OF APPROVAL No. 0198 OF 1990.

ADVISORY INFORMATION

This approval relates to the use of 'Fusilade 5' in areas of conversion from arable land or improved grassland to farm forestry and coppice woodland at a maximum rate of 3 litres product/hectare, applied in 200–500 litres water/hectare via tractor mounted sprayers only. Please note that an identical off-label approval has been issued for MAFF 02883.

NOTICE OF APPROVAL No. 0453/94

FOOD AND ENVIRONMENT PROTECTION ACT 1985
CONTROL OF PESTICIDES REGULATIONS 1986
(S.I. 1986 No. 1510):
APPROVAL FOR OFF-LABEL USE OF AN APPROVED PESTICIDE PRODUCT

This approval provides for the use of the product named below in respect of crops and situations, other than those included on the product label. Such 'off-label use' as it is known is at all times done at the user's choosing, and the commercial risk is entirely his or hers.

The conditions below are statutory. They must be complied with when the off-label use occurs. Failure to abide by the conditions of approval may constitute a breach of that approval, and a contravention of the Control of Pesticides Regulations 1986. The conditions shown below supersede any on the label *which would otherwise apply.*

Level and scope:	In exercise of the powers conferred by regulation 5 of the Control of Pesticides Regulations 1986 (SI 1986/1510) and of all other powers enabling them in that behalf, the Minister of Agriculture, Fisheries and Food and the Secretary of State, hereby jointly give full approval for the use of
Product name:	Fusilade 5 containing
Active ingredient:	125 g/l fluazifop-p-butyl
Marketed by:	ICI plc under MAFF No. 02883 subject to the conditions relating to off-label use set out below:
Date of issue:	7 March 1994
Date of expiry:	Unlimited (subject to the continuing approval of MAFF 02883)

9

Field of use:	ONLY AS AN AGRICULTURAL/ HORTICULTURAL/FORESTRY HERBICIDE
Situations:	Farm forestry
Maximum individual dose:	3 litres product/hectare
Maximum number of treatments:	Two per year
Operator protection:	(1) Engineering control of operator exposure must be used where reasonably practicable in addition to the following personal protective equipment: (a) Operators must wear suitable protective clothing (coveralls), suitable protective gloves and face protection (faceshield) when handling the concentrate. (b) Operators must wear suitable protective clothing (coveralls), suitable protective gloves when spraying and adjusting or maintaining equipment or handling contaminated surfaces. (2) However, engineering controls may replace personal protective equipment if a COSHH assessment shows they provide an equal or higher standard of protection.
Environmental protection:	Since this product is dangerous to fish or aquatic life, surface

waters or ditches must not be contaminated with chemical or used container.

Other specific restrictions:

(1) This product must only be applied if the terms of this approval, the product label and/or leaflet and any additional guidance on off-label approvals have first been read and understood.

(2) This product must not be applied via knapsack sprayers or other hand-held sprayers.

(3) This product must only be used for forestry establishment on land previously under arable cultivation or improved grassland.

9

Signed J Micklewright
(Authorised signatory)

Date 7 March 1994

Application Reference Number: COP 93/01116

THIS NOTICE OF APPROVAL IS NUMBER 0453 of 1994
THIS NOTICE OF APPROVAL SUPERSEDES NOTICE OF APPROVAL No. 0198 OF 1991.

ADVISORY INFORMATION
This approval relates to the use of 'Fusilade 5' in areas of conversion from arable land or improved grassland to farm forestry and coppice woodland at a maximum rate of 3 litres product/hectare, applied in 200–500 litres water/hectare via tractor mounted sprayers only. Please note that an identical off-label approval has been issued for MAFF 06669.

NOTICE OF APPROVAL No. 1302/94

FOOD AND ENVIRONMENT PROTECTION ACT 1985
CONTROL OF PESTICIDES REGULATIONS 1986
(S.I. 1986 No. 1510):
APPROVAL FOR OFF-LABEL USE OF AN APPROVED PESTICIDE PRODUCT

This approval provides for the use of the product named below in respect of crops and situations, other than those included on the product label. Such 'off-label use' as it is known is at all times done at the user's choosing, and the commercial risk is entirely his or hers.

The conditions below are statutory. They must be complied with when the off-label use occurs. Failure to abide by the conditions of approval may constitute a breach of that approval, and a contravention of the Control of Pesticides Regulations 1986. The conditions shown below supersede any on the label *which would otherwise apply.*

Level and scope: In exercise of the powers conferred by regulation 5 of the Control of Pesticides Regulations 1986 (SI 1986/1510) and of all other powers enabling them in that behalf, the Minister of Agriculture, Fisheries and Food and the Secretary of State, hereby jointly give full approval for the use of

Product name: Fusilade 250 EW containing

Active ingredient: 250 g/l fluazifop-p-butyl

Marketed by: Zeneca Crop Protection under MAFF NO. 06531 subject to the conditions relating to off-label use set out below:

Date of issue: 9 August 1994

Date of expiry: Unlimited (subject to the continuing approval of MAFF 06531)

Field of use:	ONLY AS AN AGRICULTURAL/ HORTICULTURAL/FORESTRY HERBICIDE
Situations:	Farm forestry
Maximum individual dose:	1.5 litres product/hectare
Maximum number of treatments:	Two per year
Operator protection:	(1) Engineering control of operator exposure must be used where reasonably practicable in addition to the following personal protective equipment: (a) Operators must wear suitable protective clothing (coveralls), suitable protective gloves and face protection (faceshield) when handling the concentrate. (b) Operators must wear suitable protective clothing (coveralls), suitable protective gloves when spraying and adjusting or maintaining equipment or handling contaminated surfaces. (2) However, engineering controls may replace personal protective equipment if a COSHH assessment shows they provide an equal or higher standard of protection.
Environmental protection:	Since this product is dangerous to fish or aquatic life, surface

waters or ditches must not be contaminated with chemical or used container.

Other specific restrictions:

(1) This product must only be applied if the terms of this approval, the product label and/or leaflet and any additional guidance on off-label approvals have first been read and understood.

(2) This product must not be applied via knapsack sprayers or other hand-held sprayers.

(3) This product must only be used for forestry establishment on land previously under arable cultivation or improved grassland.

Signed J Micklewright
(Authorised signatory)
Date 24 August 1994

Application Reference Number: COP 94/00589

THIS NOTICE OF APPROVAL IS NUMBER 1302 of 1994

ADVISORY INFORMATION

This approval relates to the use of 'Fusilade 250 EW' in areas of conversion from arable land or improved grassland to farm forestry and coppice woodland at a maximum rate of 1.5 litres product/hectare, applied in 200–500 litres water/hectare via tractor mounted sprayers only.

9.4 Metazachlor

Product

Butisan S 500 g/litre metazachlor (BASF)

Description

A residual soil acting herbicide for the control of annual dicotyledons and annual grasses in farm forestry. Most activity is pre-emergent, although some post-emergent weed control may be possible. See Table 8.

Crop tolerance

See Table 7.

Product rate

Apply 2.5 litres of product per treated hectare.

Methods of application

Pre- or post-plant (overall, band, spot or directed)

Knapsack or tractor mounted sprayers at MV.

Timing of application

Best results will be achieved from applications after planting, prior to bud burst and weed germination. Up to three applications a year may be made – effective control is 12 weeks per application.

Additional information

1. **Weed control**
   a. Apply to brash and weed-free soil, cultivated to form a firm, fine, moist tilth.
   b. Effectiveness is reduced in soils with more than 10% organic matter (peat, brash, litter, etc.) content.
   c. Rain is necessary after application for maximum weed control.
2. **Protective clothing**
   Read the product label for protective clothing and equipment requirements and check there are no items required in addition to the Forestry Commission recommendations in Section 10.
3. **Special precautions**
   a. Harmful if swallowed.
   b. Irritating to eyes and skin – avoid contact.
   c. Keeping livestock out of treated areas until any poisonous weeds such as ragwort have died and become unpalatable.

d. Do not contaminate ponds, watercourses or ditches with the chemical or used containers.

The label on there herbicide container has been designed for your protection – ALWAYS READ THE INSTRUCTIONS ON THE LABEL.

NOTICE OF APPROVAL No. 0227/94

FOOD AND ENVIRONMENT PROTECTION ACT 1985 CONTROL OF PESTICIDES REGULATIONS 1986 (S.I. 1986 No. 1510): APPROVAL FOR OFF-LABEL USE OF AN APPROVED PESTICIDE PRODUCT

This approval provides for the use of the product named below in respect of crops and situations, other than those included on the product label. Such 'off-label use' as it is known is at all times done at the user's choosing, and the commercial risk is entirely his or hers.

The conditions below are statutory. They must be complied with when the off-label use occurs. Failure to abide by the conditions of approval may constitute a breach of that approval, and a contravention of the Control of Pesticides Regulations 1986. The conditions shown below supersede any on the label *which would otherwise apply.*

Level and scope:	In exercise of the powers conferred by regulation 5 of the Control of Pesticides Regulations 1986 (SI 1986/1510) and of all other powers enabling them in that behalf, the Minister of Agriculture, Fisheries and Food and the Secretary of State, hereby jointly give full approval for the use of
Product name:	Butisan S containing
Active ingredient:	500 g/l metazachlor
Marketed by:	BASF plc under MAFF No. 00357 subject to the conditions relating to off-label use set out below:
Date of issue:	27 January 1994
Date of expiry:	Unlimited (subject to the continuing approval of MAFF 00357)

Field of use:	ONLY AS AN AGRICULTURAL/ FORESTRY HERBICIDE
Situations:	Farm forestry
Maximum individual dose:	2.5 litres product/hectare
Maximum number of treatments:	Three per year
Operator protection:	(1) Engineering control of operator exposure must be used where reasonably practicable in addition to the following personal protective equipment: (a) Operators must wear suitable protective gloves and face protection (faceshield) when handling the concentrate. (b) Operators must wear suitable protective gloves when handling the spray boom or adjusting nozzles. (2) However, engineering controls may replace personal protective equipment if a COSHH assessment shows they provide an equal or higher standard of protection.
Environmental protection:	Since this product is dangerous to fish or aquatic life, surface waters or ditches must not be contaminated with chemical or used container.
Other specific restrictions:	(1) This product must only be applied if the terms of this approval, the product label

and/or leaflet and any additional guidance on off-label approvals have first been read and understood.

(2) Livestock must be kept out of treated areas until foliage of any poisonous weeds, such as ragwort, have died and become unpalatable.

(3) Crops grown within the treated area of woodland must not be used for human or animal consumption.

(4) This product must only be used for forestry establishment on land previously under arable cultivation or improved grassland.

Signed J Micklewright
(Authorised signatory)

Date 27 January 1994

Application Reference Number: COP 93/01118

THIS NOTICE OF APPROVAL IS NUMBER 0227 of 1994
THIS NOTICE OF APPROVAL SUPERSEDES NOTICE OF APPROVAL No. 0191 OF 1990.

ADVISORY INFORMATION

This approval relates to the use of 'Butisan S' in areas of conversion from arable land or improved grassland to farm forestry and coppice woodland. Application may be made by conventional tractor mounted ground spray equipment or knapsack sprayers.

9.5 Pendimethalin

Product

Stomp 400 400 g/litre pendimethalin (Cyanamid)

Description

A soil acting residual herbicide for the control of annual grasses and annual dicotyledons in farm forestry. See Table 8.

Crop tolerance

See Table 7.

Product rate

Apply 5.0 litres of product per treated hectare.

Methods of application

Pre- or post-plant (overall, band, spot or directed)
Knapsack or tractor mounted sprayers at LV–MV.

Timing of application

Best results will be achieved from applications immediately after planting, prior to bud burst or weed germination.
One application a year is allowed.

Additional information

1. **Weed control**
a. For best results, apply to a brash and weed-free site, cultivated to achieve a firm, moist tilth.
b. Do not use in soils with greater than 10% organic matter (peat, brash, litter, etc.) content.
c. Effectiveness is reduced by prolonged dry weather after treatment.
2. **Protective clothing**
Read the product label for protective clothing and equipment requirements and check there are no items required in addition to the Forestry Commission recommendations in Section 10.
3. **Special precautions**
a. Dangerous to fish – avoid contamination of ponds, watercourses or ditches with the chemical or used containers.

The label on there herbicide container has been designed for your protection – ALWAYS READ THE INSTRUCTIONS ON THE LABEL.

NOTICE OF APPROVAL No. 0226/94

FOOD AND ENVIRONMENT PROTECTION ACT 1985
CONTROL OF PESTICIDES REGULATIONS 1986
(S.I. 1986 No. 1510):
APPROVAL FOR OFF-LABEL USE OF AN APPROVED PESTICIDE PRODUCT

This approval provides for the use of the product named below in respect of crops and situations, other than those included on the product label. Such 'off-label use' as it is known is at all times done at the user's choosing, and the commercial risk is entirely his or hers.

The conditions below are statutory. They must be complied with when the off-label use occurs. Failure to abide by the conditions of approval may constitute a breach of that approval, and a contravention of the Control of Pesticides Regulations 1986. The conditions shown below supersede any on the label *which would otherwise apply*.

Level and scope:	In exercise of the powers conferred by regulation 5 of the Control of Pesticides Regulations 1986 (SI 1986/1510) and of all other powers enabling them in that behalf, the Minister of Agriculture, Fisheries and Food and the Secretary of State, hereby jointly give full approval for the use of
Product name:	Stomp 400 SC containing
Active ingredient:	400 g/l pendimethalin
Marketed by:	Cyanamid of Great Britain Ltd under MAFF No. 04183 subject to the conditions relating to off-label use set out below:
Date of issue:	27 January 1994
Date of expiry:	Unlimited (subject to the continuing approval of MAFF 04183)

9

Field of use:	ONLY AS AN AGRICULTURAL/ FORESTRY HERBICIDE
Situations:	Farm forestry
Maximum individual dose:	5 litres product/hectare
Maximum number of treatments:	One per year
Environmental protection:	Since this product is dangerous to fish or aquatic life, surface waters or ditches must not be contaminated with chemical or used container.
Other specific restrictions:	(1) This product must only be applied if the terms of this approval, the product label and/or leaflet and any additional guidance on off-label approvals have first been read and understood. (2) Crops grown within the treated area of woodland must not be used for human or animal consumption. (3) This product must only be used for forestry establishment on land previously under arable cultivation or improved grassland.

Signed J Micklewright
(Authorised signatory)

Date 27 January 1994

Application Reference Number: COP 93/01112

THIS NOTICE OF APPROVAL IS NUMBER 0226 of 1994

THIS NOTICE OF APPROVAL SUPERSEDES NOTICE OF APPROVAL No. 0377 OF 1991.

ADVISORY INFORMATION

This approval relates to the use of 'Stomp 400 SC ' in areas of conversion from arable land or improved grassland to farm forestry and coppice woodland at a maximum rate of 5 litres product/hectare applied in 200–400 litres water/hectare via tractor mounted or knapsack sprayer.

9.6 Propaquizafop

Products

Falcon	100 g/litre propaquizafop (Ciba Geigy, marketed by Cyanamid)
Shogun 100 EC	100 g/litre propaquizafop (Ciba)

Description

A foliar acting translocated herbicide for the control of annual and perennial grasses in farm forestry. This herbicide will have little effect on herbaceous broadleaved weeds. See Table 8.

Crop tolerance

See Table 7.

Product rate

Apply at 0.7–1.0 litres of product per treated hectare for volunteer cereals and black grass, 1.2 1/ha for rye-grasses and 1.5 l/ha for common couch. Maximum total application rate is 2.0 litres of product per hectare per year.

Methods of application

Pre- or post-plant (overall, band, spot or directed)

Knapsack or tractor mounted sprayers at MV (200–250 l/ha).

Timing of application

Timing of application is dependent on the growth stage of the target weed, and the crop present – see Tables 7 and 8.

Additional information

1. **Weed control**

a. Control will be most rapidly achieved in warm, moist conditions when weeds are actively growing.

b. Use the higher rate specified for individual weed species if the weeds are overwintered, beyond the specified optimum growth stage for control, for severe infestations and in poor growing conditions for weeds such as cool temperatures and dry soil.

c. Some strains of black grass may be resistant to applications of propaquizafop.

2. **Protective clothing**

Read the product label for protective clothing and equipment requirements and check there are no items required in addition to the Forestry Commission recommendations in Section 10.

3. Special precautions

a. Harmful to fish and aquatic life – do not contaminate ponds, watercourses or ditches with the chemical or the used container.

The label on the herbicide container has been designed for your protection – ALWAYS READ THE INSTRUCTIONS ON THE LABEL.

10 Protective clothing and personal equipment

10.1 General

The appropriate protective equipment, listed in Table 10, should be made available on a personal basis to all users of herbicides, including those handling herbicide containers.

Always read the product label in case there is need for additional items of protective clothing or equipment to that recommended in Table 10.

All protective equipment should be kept clean and in good repair. Any damaged item should be replaced promptly.

To avoid contamination of personal clothing or skin it is essential that the contaminated outside of the clothing does not come into contact with the clean inside, during transit or when putting on or removing protective clothing and equipment.

The order for removing clothing and equipment is:
first wash gloves and then remove in the following order,

- face shield
- filtering facepiece respirator
- impervious suit (either one or two piece)
- wellington boots
- gloves (after washing again to remove any contamination picked up during clothing removal).

Clean clothing should be put on in reverse order without the necessity of washing gloves, except if a new respirator is used, in which case it should be put on first.

At the end of the day all clothing and equipment should be thoroughly washed down, dried and left to hang overnight.

At the end of the spraying season, or more frequently if necessary, items should be thoroughly washed and dried either at the forest store or by a reputable cleaning company. Items must never be washed with normal domestic washing. Cleaned items should be hung up and stored in a cool dry place away from direct sunlight, vermin and pesticides.

Contaminated items of protective clothing and equipment must not be worn or carried inside the cab of any vehicle.

10.2 List of recommended products and suppliers

Equipment	**Recommendations**	**Suppliers** (see footnote)
Wellington boots	Dunlop Safety 8807 or 8808 with steel midsole	Greenham Tool Co. Ltd. 671 London Road Isleworth Middlesex TW7 4EX
Shoe chains	Rudd Shoe Chains Size 1: Shoe sizes under 5 Size 2: Shoes sizes 5-9 Size 3: Shoe sizes over 9	Rudd Chains Ltd. 1-3 Belmont Road Whitstable Kent CT5 1QT
Impervious suit (trousers & jacket with hood)	Cascade suits in small, medium & large sizes	Edward MacBean and Co Ltd. 1-7 Napier Place Wardpark North Cumbernauld Glasgow G68 0LL *For FC staff:* Blairadam Clothing Store
Gloves	Edmont Solvex NX37-175 Length 30 cm sizes 7-11 It is recommended that two pairs of gloves are issued to each operator	Safety Specialist Ltd. Unit 8 Field House Industrial Estate Peter Street Sheffield S4 7SF
Cotton liner gloves (for use with the Forestry Spot Gun)	Mens Cotton Stockinette Gloves, knitted wrist: Code 304111	Greenham Tool Co. Ltd. 671 London Road Isleworth Middlesex TW7 4EX
Face shield	James North No. FS 1318 BW Shield depth 10"	Greenham Tool Co. Ltd. 671 London Road Isleworth Middlesex TW7 4EX
Face shield for use with safety helmet for drivers of tractors without cabs	FC H417 Aluminium frame for Safety Helmet FL 8 PC (20 cm) clear polycarbonate screen wide flare	Protective Safety Ltd. Great George Street Wigan Greater Manchester WN3 4DE
Filtering face-piece respirator complying to FFP2S	3 M's 8810 Generally requires replacing after 2 hours use on a misty day or every 4 hours on a dry day	Herts Packaging Co. Ltd. 29 Mill Lane Welwyn Herts AL6 9EU

Equipment	Recommendations	Suppliers (see footnote)
Personal hygiene		
Barrier cream Options should be available to allow for varying skin conditions	Rozalex Wet Guard available in 450 ml containers	Lever International Ltd. PO Box 208 Lever House St James Road Kingston-on-Thames Surrey KT1 2BB
Skin cleanser for use with water	Arrow Chemical Tuffstuff available in 1 litre dispenser bottle or 5, 10 and 20 litre containers	The Arrow Chemical Group of Companies PO Box 3 Stanhope Road Swadlincote Near Burton-on-Trent Staffs DE11 9BE
Paper towels	Kimwipes Steel Blue 10" rolls, Code 7148 In a case of 24 rolls	Kimberley Clark Ltd. Industrial Division Larkfield Kent ME20 7PS (Private sector: purchase locally from any industrial clothing supplier)

Note:
This list is included as a guide to sources of supply and is NOT a comprehensive compilation of suppliers of recommended products. The omission of names of other possible suppliers does not imply that their services are unsatisfactory.

10.3 Cleaning recommendations

Any item of protective clothing or equipment which becomes grossly contaminated should be immediately rinsed thoroughly with water.

Wellington boots

At the end of each day wash down outsides of boots in a mild dilute solution of detergent and rinse in clean water. Allow both inside and outside to dry.

Gloves

Gloves should be washed:

- at the end of each work period, before removing protective clothing;
- after removing protective clothing;
- at the end of each working day.

Face shield

Wipe down with a mild dilute solution of detergent and allow to dry.

Impervious coverall (Cascade jacket and trousers)

At the end of each day wash down with a mild dilute solution of a household washing detergent (not washing up liquid), and rinse in clean water. Allow both the inside and outside to dry.

Do not use high pressure hoses to wash down the suits.

When Cascade suits are in use they should be handled with care as rough treatment will reduce the liquid repellent properties.

To test the repellent properties of the Cascade material, paper towelling should be placed inside the suit and water applied, without pressure, together with a light rubbing action to the outside of the garment for 5–10 seconds. If the paper becomes wet the suit will have lost its repellent properties and should be replaced.

Table 10 Protective clothing and equipment recommendations

Operation	Application equipment	Wellington boots	Impervious coverall		Hood	Gloves	Face shield	Filtering facepiece respirator	Ear defenders	Notes
			Trousers	*Jacket*						
Handling concentrate, mixing and filling (granules and liquid)	All types	E	E	E	E	E	E	D	–	
Application of granules of crystals	Gravity	E	E	D	D	E	D	D	–	
Medium volume (MV) and low volume (LV) spraying	Knapsack, nozzle height less than 1 metre	E	E	E	D	E	D	D	–	
	Knapsack, nozzle height greater than 1 metre	E	E	E	E	E	E	E	–	
	Forestry spot gun nozzle height less than 1 metre	E	E	D	D	E	D	–	–	(1)

Table 10 Protective clothing and equipment recommendations – *continued*

Operation	Application equipment	Wellington boots	Impervious coverall		Hood	Gloves	Face shield	Filtering facepiece respirator	Ear defenders	Notes
			Trousers	*Jacket*						
	Forestry spot gun nozzle height greater than 1 metre	E	E	E	E	E	E	E	–	(1) (3) (4)
Controlled droplet applicator (band/ overall)	Microfit Herbi (low speed rotary atomiser)	E	E	E	D	E	D	D	–	(3) (4)
Controlled droplet applicator (incremental/ placed)	Ulva 8/Ulva+ (high speed rotary atomiser)	E	E	E	E	E	E	E	–	(3) (4)
Direct applicator	Weedwiper	E	E	D	D	E	D	–	–	(3)(4)
Stem treatment tree injection	Spot gun	E	E	E	D	E	D	–	–	(3) (4)
Brushcutter stump treatment attachment	ENSO	E	E	D	D	E	D	D	E	(3) (4)

10

10.4

Table 10 Protective clothing and equipment recommendations – *continued*

Operation	Application equipment	Wellington boots	Impervious coverall		Hood	Gloves	Face shield	Filtering facepiece respirator	Ear defenders	Notes
			Trousers	*Jacket*						
Tractor mounted sprayers (MV, LV, VLV)	Tractor with fully enclosed cab (activated carbon filters must be fitted).	–	–	–	–	–	–	–	–	(2)
	Tractor without cab	E	E	E	E	E	D	–	E	(2)

Notes:

(1) For operator comfort, cotton liner gloves may be worn during application.

(2) When repairing or maintaining the sprayer the clothing requirements are as for handling, mixing and filling.

(3) When using hand-held applicators to apply products containing 2,4-D, operators must wear suitable protective boots, trousers, jacket, hood, gloves and face shield.

(4) When using hand-held applicators to apply products containing atrazine, operators must wear suitable protective boots, trousers, jacket, hood, gloves, face shield and filtering facepiece respirator.

E = essential: either advised under Control of Pesticides Regulations 1986 and Control of Substances Hazardous to Health Regulations 1988 or considered necessary in relation to working condition.

D = discretionary: these items are not usually required but should be supplied on request to the operator or when noticeable exposure to a herbicide may arise through operations in unusual circumstances.

Recommendation for protective clothing and equipment requirements for use with application equipment not listed should be obtained from: Technical Development Branch, Ae Village, Dumfries DG1 1QB. Telephone 01387 860264.

11 Equipment

List of contents *page*

11.1 Disclaimer 215
11.2 Volume rate categories 216
11.3 Calibration 217
11.3.1 Principles 217
11.3.2 Calculation 217
11.3.3 Calibration equations 218
11.4 Nozzles 221
11.5 Applicators for liquid herbicides 222
11.5.1 Cooper Pegler Series 2000 CP15 forestry model knapsack sprayer 222
11.5.2 Dribble bar 235
11.5.3 Fox Motori Electra F110 TS knapsack sprayer 237
11.5.4 Tecnoma T18P knapsack sprayer 240
11.5.5 Forestry spot gun 241
11.5.6 Microfit Herbi low speed rotary atomiser 243
11.5.7 ULVA 8/ULVA + high speed rotary atomiser 248
11.5.8 Handheld direct applicator Weedwiper Mini 255
11.5.9 Ulvaforest tractor-mounted sprayer 257
11.6 Applicators for spot application of granules 258
11.6.1 Pepperpots 258
11.6.2 Tree Mate Kerb granule applicator 259
11.7 Tree injection applicator – Forestry spot gun 260
11.8 Enso brush cutter stump treatment attachment 261
11.9 Pesticide transit box 262
11.10 Output guides – herbicide application 262
11.11 Output guide – overall or band application 263
11.12 Output guide – spot application 271

11.1 Disclaimer

The applicators detailed in this section are those which are currently recommended for use within the Forestry Commission; this does not imply that other manufacturers' applicators with similar capabilities and features are not suitable for applying the herbicides detailed in this Field Book.

11.2 Volume rate categories

High volume (HV) – over 700 litres per hectare

Not recommended for herbicide application except for cut stump spot treatment where high volume rate is recommended for optimum results. In all situations high volume application can result in waste and ground contamination due to run off. Better results with practically no run off can usually be obtained by using lower volumes per hectare.

Applicators

Knapsack (cut stump application only).

Forestry spot gun (cut stump application only).

Medium volume (MV) – 200 to 700 litres per hectare

These rates give a good overall cover in most situations.

Applicators

Tractor mounted boom sprayer.

Knapsacks.

Low volume (LV) – 50 to 200 litres per hectare

Good coverage is achieved in pressure controlled sprayers by the use of precisely engineered hydraulic nozzles.

Applicators

Tractor mounted boom sprayers fitted with LV nozzles.

Knapsacks with VLV nozzles.

Forestry spot gun (for use on grasses and herbaceous broadleaved weeds and for woody weed foliar application).

Very low volume (VLV) – 10 to 50 litres per hectare

Controlled droplet applicators (rotary atomisers) or precisely engineered hydraulic nozzles must be used at these volumes to obtain a good droplet size spectrum and good droplet dispersal.

Applicators

Low speed rotary atomisers.

High speed rotary atomisers.

Forestry spot gun (for use on grasses and herbaceous broadleaved weeds).

Ultra low volume (ULV) – under 10 litres per hectare

Controlled droplet application (rotary atomiser) must be used to obtain adequate control of droplet size and droplet dispersal.

Applicator
High speed rotary atomisers.

11.3 Calibration

11.3.1 Principles

It is a requirement of the Control of Pesticides Regulations 1986 that maximum application rates are not exceeded. Overdosing with herbicide is expensive and places the operator, the environment and crop trees at unnecessary risk. Underdosing can also be expensive if failure to control the weeds results in the need for subsequent operations. It is therefore desirable to achieve the target rate of application.

11.3.2 Calculation

To achieve the target rate of application distributed evenly over the target area it is necessary to determine the four parameters indicated below.

Volume rate (i.e. the quantity of spray mixture (diluent plus herbicide) applied in litres per treated hectare)

From the Method of application paragraph of the selected herbicide, determine the appropriate applicator and therefore the required volume rate.

Any volume rate within the range (i.e. MV 200–700) can be selected by varying the amount of diluent but keeping the rate at which the herbicide is applied constant.

Throughout this Field Book the rate of herbicide to be applied is given in litres or kilograms per **TREATED** hectare.

Speed (walking speed or tractor speed)

Walking speed
Measure the walking speed (in metres per minute), which can be sustained on site with the applicator full, and wearing the necessary protective clothing. Deviation from this speed will lead to over or underdosing.
Ground conditions may dictate the walking speed.

Tractor speed
This is determined by the roughness of the terrain.

Swathe width

Determine the swathe width (or spot diameter). Note that the height at which the nozzle is held has to be increased or decreased to take into account changes in vegetation height in order to maintain the measured swathe width at the average height of the target and avoid over or underdosing.

Nozzle output

Measure the nozzle output over a set period and convert to millilitres per minute. Note that changes in air temperature in the course of the day can alter the viscosity of the herbicide, requiring nozzle output to be remeasured. Of the four parameters, nozzle output is generally the most appropriate to vary, either by adjusting the pressure or changing nozzles.

11.3.3 Calibration equations for calculating various parameters applicable to different types of applicators

Equation No.

1.
$$\text{Nozzle output (ml/mm)} = \frac{\text{Walking speed (metres/min)} \times \text{Volume rate (litres/ha)} \times \text{Swathe width (metres)}}{10}$$

2.
$$\text{Walking speed (metres/min)} = \frac{\text{Nozzle output (ml/min)} \times 10}{\text{Volume rate (litres/ha)} \times \text{Swathe width (metres)}}$$

3.
$$\text{Volume rate (litres/ha)} = \frac{\text{Nozzle output (ml/min)} \times 10}{\text{Walking speed (metres/min)} \times \text{Swathe width (metres)}}$$

4.
$$\text{Distance covered when band spraying (metres)} = \frac{\text{Total volume of sprayer (litres)} \times 10\,000}{\text{Volume rate (litres/ha)} \times \text{Swathe width (metres)}}$$

5. Volume of liquid herbicide product required when band spraying (litres)

$$\text{Volume of liquid herbicide product required when band spraying (litres)} = \frac{\text{Area of plantation (ha)} \times \text{Product rate (litres/ha)} \times \text{Swathe width (metres)}}{\text{Row spacing (metres)}}$$

The area of plantation actually treated when band spraying is presented in ready reckoner form in Table 12.

6. Volume per spot (millilitres)

$$\text{Volume per spot (millilitres)} = \frac{\text{Volume rate (litres/ha)} \times \text{Area of spot } (m^2)}{10}$$

7. Volume rate for spot treatment (litres/ha)

$$\text{Volume rate for spot treatment (litres/ha)} = \frac{\text{Volume per spot (ml)} \times \text{Number of trees } (ha^{-1})}{1000}$$

8. Number of trees sprayed with spot treatment

$$\text{Number of trees sprayed with spot treatment} = \frac{\text{Volume of liquid in sprayer (litres)}}{\text{Volume per tree (litres)}}$$

9. Distance covered by a tank full of spray when spot spraying (metres)

$$\text{Distance covered by a tank full of spray when spot spraying (metres)} = \frac{\text{Volume of liquid in sprayer (litres)} \times 1000 \times \text{Tree spacing in rows (metres)}}{\text{Volume rate (litres/ha)} \times \text{Area of spot } (m^2)}$$

10. Quantity of herbicide product required (ml) per applicator/container

$$\frac{\text{Applicator/container capacity (litres)} \times \text{Product rate (litres/ha)} \times 1000}{\text{Volume rate (litres/ha)}}$$

Table 11 **The proportion of 1 hectare of plantation treated during spot applications**

Spot diameter *(m)*	Spot area *(m²)*	Tree spacing (m) (assumes fully stocked square planting)							
		1.7	*1.8*	*1.9*	*2.0*	*2.1*	*2.2*	*2.3*	*3.0*
1.5	1.77	.61	.55	.49	.44	.40	.37	.33	.19
1.4	1.54	.53	.48	.43	.39	.35	.32	.29	.17
1.3	1.33	.46	.41	.37	.33	.30	.27	.25	.14
1.2	1.13	.39	.35	.31	.28	.26	.23	.21	.13
1.1	.95	.33	.29	.26	.24	.22	.20	.18	.11
1.0	.79	.27	.24	.22	.20	.18	.16	.15	.09

Table 12 **The proportion of 1 hectare of plantation treated during band applications**

Width of swathe *(m)*	Row spacing (m)							
	1.7	*1.8*	*1.9*	*2.0*	*2.1*	*2.2*	*2.3*	*3.0*
1.5	.88	.83	.79	.75	.71	.68	.65	.50
1.4	.82	.78	.74	.70	.67	.64	.61	.47
1.3	.76	.72	.68	.65	.62	.59	.57	.43
1.2	.71	.67	.63	.60	.57	.55	.52	.40
1.1	.65	.61	.58	.55	.52	.50	.48	.37
1.0	.59	.56	.53	.50	.48	.45	.43	.33

11.4 Nozzles

General information

With hydraulic nozzles, increasing the pressure will increase the nozzle output, reduce the size of the droplet and possible widen the swathe width or spot diameter. Decreasing the pressure has the opposite effects.

All nozzles should be inspected regularly for wear and damage which can alter their outputs, droplet size and distribution characteristics.

Damage to the spinning discs of the rotary atomisers will affect droplet size.

Any nozzle with visual signs of wear or damage should be replaced immediately.

Nozzle manufacturers have different conventions of defining pressures: bar, pounds per square inch (p.s.i.) and kilopascals (kPa). The units of conversion are:

1 bar = 14.5 p.s.i. = 100 kPA

Hydraulic nozzles (fitted to the knapsack and forestry spot gun)

The three types of hydraulic nozzles recommended in this Field Book are:

a. **Floodjet** (also known as anvil or deflector) **nozzles** produce a wide angled flat spray with an even distribution over the swathe width. At low pressure these nozzles produce large droplets with little risk of drift.

   Flat even spray fan nozzles which also produce an even distribution over the swathe width are available in a range of angles and produce a similar pattern to the floodjet.

   These work at a slightly higher pressure thus producing smaller droplets with a higher risk of drift.

   Normal flat fan nozzles are unacceptable for forestry as they do not produce an even distribution over the swathe width and therefore have to be overlapped to produce an even distribution over the treated area.

b. **Solid cone nozzles** distribute droplets over the entire area of a spot, where the spot diameter is dependent on the angle of the nozzle and the height at which the nozzle is held.

c. **Solid stream nozzles** deliver the spray solution as a solid stream and are used for stem injection, stump treatments and application of dicamba in a narrow band between crop rows.

Feed nozzles (fitted to the Herbi, ULVA and Ulvaforest)

Varying orifice sizes regulate the flow of the spray solution to the spinning disc.

For further information on nozzles refer to the *Nozzle selection handbook* published by the British Crop Protection Council – available from BCPC Sales, Bear Farm, Binfield, Bracknell, Berkshire RG12 5QE.

11.5 Applicators for liquid herbicides

11.5.1 Cooper Pegler Series 2000 CP15 forestry model knapsack sprayer with pressure control valve or spray management valve

Uses

Cut stump application at high volumes.
General herbicide application at medium and low volumes.

Supplier

Cooper Pegler Spraying Technology
North Seaton Industrial Estate
Ashington
Northumberland
NE63 0XA
Telephone: 01670 522225

Description

CP15 Forestry Model 15 litre knapsack sprayer, complete with pressure gauge and adaptor, pressure control valve and adaptor, and hose assembly. Ref. No. PA 1104.

Accessories

Spray shield 30 mm (12") complete with Green Polijet
Ref. No. SA 04 630
Spray shield 35 mm (15") complete with Green Polijet
Ref. No. SA 04 631
Politec Tree Guard with two nozzles. Ref. No. SA 04 637

Table 13

Nozzle-output data – CP15 knapsack sprayer

a. Floodjet

	Nozzle type	Nozzle output @ 0.7 bar (ml/min with water)	Nozzle output @ 1 bar (ml/min with water)
(Volume rate	Polijet		
– medium volume)	Red	2000	2400
	Blue	1500	1800
	Green (Note 1)	1000	1200
(Volume rate	Very low volume		
– low volume)	(Notes 1, 2)		
	AN 2.0 Blue	730	920
	AN 1.0 Orange	360	460

Notes:

1. Also for use in the spray shield.
2. Although termed very low volume nozzles these apply at the low volume rate.

b. Solid cone with green swirl core

	Nozzle type	Nozzle output @ 0.7 bar (ml/min with water)	Nozzle output @ 1 bar (ml/min with water)
(Volume rate	5 Blue	1000	1240
– medium volume)	6 Yellow	1320	1670

c. Solid cone without swirl core

	Nozzle type	Nossle output @ 1.25 bar	Nozzle output @ 2.5 bar
	Lurmark Grey	280 ml/min	440 ml/min

11

Tools required for maintenance and calibration

Pliers
Large screwdriver
Medium screwdriver
Adjustable spanner
Roll of PTFE plumber's tape
Plastic bucket
Large funnel
Metric graduated measure
Absorbent paper
Watch
Tape measure
Spare nozzles

Calibration

Overall or band application

1. Determine the walking speed which an operator can sustain over the site with a full applicator and wearing the necessary protective clothing.
2. Select volume rate from the Method of application paragraph of the selected herbicide.
3. Measure the swathe width to be treated, (from a test application using water).
4. Calculate nozzle output using the following equation.

$$\text{Nozzle output (ml/min)} = \frac{\text{Walking speed (metres/min)} \times \text{Volume rate (litres/ha)} \times \text{Swathe width (metres)}}{10}$$

5. Select and fit a nozzle, referring to nozzle data table and manufacturer's data, which gives the closest but greater output to your calculated figure at either 0.7 or 1.0 bar depending on prevailing weather conditions.
6. Fill the applicator with water and fully pressurise.
7. Set the pressure gauge by adjusting the screw of the pressure control valve to an initial pressure of 0.7 or 1.0 bar.
8. Spray into a graduated measure for a set time and calculate nozzle output (ml/min).
9. Compare the actual nozzle output with the required nozzle output. If there is a difference adjust the pressure and re-measure output until the actual and required nozzle outputs agree.

10. When outputs agree note the pressure gauge reading for that sprayer. Pressure should be checked two or three times during the day using the same pressure gauge.
11. Confirm swathe width – spray water over paper towling spread over vegetation and measure required nozzle height to achieve target swathe width. This height must be maintained to achieve accurate application. A measured piece of light chain hung from the end of the lance can be used as an aid.
12. Calculate the required volume of herbicide concentrate per applicator/container using equation 10, Section 11.3.3.
13. Having filled the applicator in the approved manner with herbicide diluted to the required volume, check nozzle output. If different re-calibrate.

Calibration procedure when using spray management valves

1. Determine the walking speed which an operator can sustain over the site with a full applicator and wearing the necessary protective clothing.
2. Select the correct volume rate from the product label.
3. Use Table 14 to determine the options of nozzle tip available – enter the table by walking speed and volume rate.
4. Select and fit the nozzle to the lance.
5. Fill the knapsack with water and fully pressurise.
6. Open the trigger and direct the spray away from a graduated measure for 15 seconds.
7. Transfer the spray into a graduated measure for a set time and measure nozzle output.
8. Due to manufacturing tolerances, nozzles of the same type may not give exactly the same outputs. Use the following equation to check that the volume rate is within the allowed rates from the product label.

$$\text{Volume rate} = \frac{\text{nozzle output (ml/min)} \times 10}{\text{walking speed (metres/min)} \times \text{swathe width (metres)}}$$

9. If the volume rate is outwith the product label recommendations a different nozzle size will have to be selected and the volume rate checked once more.

10. If the volume rate is within label recommendations then the quantity of herbicide required, in litres per applicator, may be calculated using the following equation.

$$\text{Herbicide product required (ml)} = \frac{\text{Applicator capacity (litres)} \times \text{Product rate (litres/ha)}}{\text{Volume rate (litres/ha)}} \times 1000$$

Table 14 Nozzle output data, when using a knapsack sprayer fitted with a spray management valve

Nozzle tip	Nozzle tip (ml/min at 1.0 bar)	Walking speed (metres per minute) 20	25	30	35	40	45	50	55	60
		Volume rate (litres per hectare)								
Pink AN 0.5	230	115	92	77	66	58	51	46	42	38
Light brown AN 0.75	350	175	140	117	100	88	78	70	64	58
Orange AN 1.0	460	230	184	153	131	115	102	92	84	77
Red AN 1.5	690	345	276	230	197	172	153	138	125	115
Cambridge blue AN 2.0	920	460	368	307	263	230	204	184	167	153
Yellow AN 2.5	1150	–	460	383	329	288	256	230	209	192
Lime green AN 3.0	1390	–	–	463	397	348	309	278	253	232
Moss green AN 4.0	1850	–	–	–	–	462	411	370	336	308
Royal blue AN 5.0	2310	–	–	–	–	–	–	462	420	385

Notes:

1. Equivalent ICI Polijet/Lurmark nozzle tips:
   ICI Red Polijet 2350 ml/min = Lurmark AN 5.0 Royal Blue 2210 ml/min
   ICI Blue Polijet 1670 ml/min = Lurmark AN 4.0 Moss Green 1850 ml/min = Lurmark AN 3.0 Lime Green 1390 ml/min
   ICI Green Polijet 910 ml/min = Lurmark AN 2.0 Cambridge Blue 920 ml/min
2. For different walking speed, volume rate can be calculated using the following equation:

$$\text{Volume rate (litres/ha)} = \frac{\text{Nozzle output (ml/min)} \times 10}{\text{Walking speed (metres/min)} \times \text{Swathe width (metres)}}$$

For band or overall applications with a knapsack sprayer, users may find the step by step summary of the calibration process that follows easier to use. Permission is granted by the Forestry Commission to copy the following calibration guide for personal use.

Calibration of a knapsack sprayer with a pressure control valve (PCV)

A Choose herbicide – refer to Field Book 8 and product labels

Note: Check whether use of knapsack sprayer is permissible and at what volume rate

B Decide volume rate (from product label) (1) [ ] l/ha

C Check protective clothing – Field Book 8 Section 10.4, Table 10.

D Determine the actual walking speed

The operator, wearing correct protective clothing and spraying water walks in a straight line for a period of 30 seconds or 1 minute. The distance is measured and the walking speed in metres/minute determined.

Walking speed (2) [ ] metres/min

E Decide desirable swathe width

Width chosen will depend on weed control required. Make a test application as described in step I to determine if desired swathe width is practical.

(3) [ ] metres

F Calculate nozzle output

Nozzle output (ml/min) = (Walking speed × Volume rate × Swathe width) ÷ 10

(4) [ ] ml/min = [(1) [ ] × (2) [ ] × (3) [ ]] ÷ 10

G Taking weather conditions into account, select a nozzle by referring to the manufacture's data. In windy conditions a pressure of 0.7 bar should be adopted in preference to 1.0 bar to reduce spray drift.

H Confirm nozzle output

Pressurise sprayer and adjust pressure control valve to obtain a pressure of either 0.7 or 1.0 bar. Spray at this pressure for a set time into a graduated measure. Adjust pressure and output using PCV to obtain calculated output. Note the pressure gauge reading for *that* sprayer. Pressure should be checked two or three times during the day.

I Confirm swathe width

Spray water over paper towelling spread over target vegetation and measure required nozzle height to achieve required swathe width. This height must be maintained to achieve accurate application. If this is uncomfortable, consider changing the nozzle although such a change will mean re-calculation of the nozzle output.

J Obtain product rate (from product label)

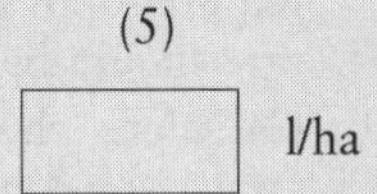

K Applicator tank volume (*Note:* May be part filled)

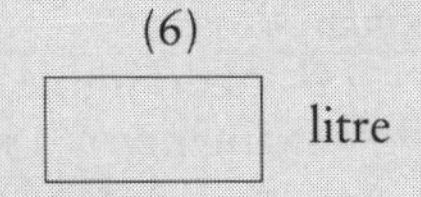

L Calculate volume of pesticide product for each tankful

Applicator tank volume × Product rate × 1000/ Volume rate = ml/tank

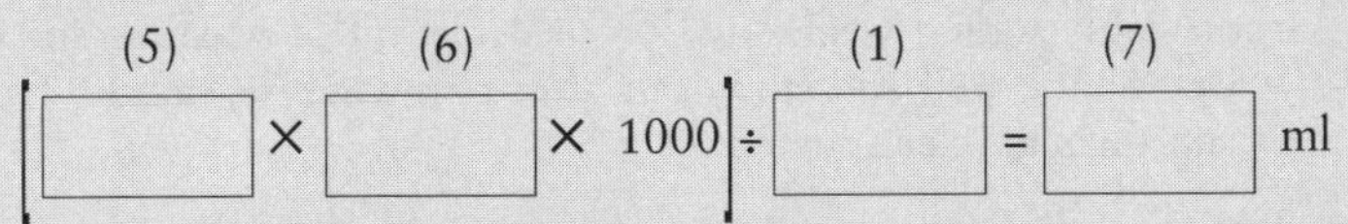

M Having filled the applicator in the approved manner with herbicide diluted to the required volume, check nozzle output. If different, re-calibrate.

11

Spot application using knapsack sprayers (Note – The forestry spot gun is preferred equipment – see Section 11.5.5)

1. Select volume rate from Method of application paragraph of the selected herbicide.
2. Measure spot diameter (from a test application using water with the lance held at a comfortable length).
3. Calculate volume per spot using the following equation:

$$\text{Volume per spot (millilitres)} = \frac{\text{Volume rate (litres/ha)} \times \text{Area of spot } (m^2)}{10}$$

4. From the nozzle output data select and fit to the knapsack the nozzle which gives the nearest (but greater) output than the required nozzle output.
5. Fill the applicator with water and fully pressurise.
6. Time yourself spraying 10 spots.
7. Multiply calculated volume/spot (ml) (from step 3) by 10.
8. Adjust knapsack pressure control valve to deliver the calculated volume for 10 spots in the time it took to spray 10 spots.
9. Practise application into a graduated measure until the desired volume per spot is achieved; then over a dry surface to achieve the required spot size. If the application time is not practical the volume rate should be adjusted until a realistic time is achieved.
10. Calculate the required volume of herbicide concentrate per applicator container using equation 10.
11. Having filled the applicator in the approved manner with herbicide diluted to the required volume, check nozzle output. If different re-calibrate.

Droplet size

If undesirable drift is occurring use a larger droplet size by changing to a nozzle of the same type but with a larger hole and use at a lower pressure to obtain the same nozzle output. If necessary, adjust the working height of the nozzle to obtain the same width of treatment.

Cleaning

Spray out all dilute pesticide safely (see Section 3.9). Wash thoroughly with a weak solution of detergent or washing soda and water. Shake well and spray out after removing the nozzle. Finally, pump through clean water.

For spot applications with a knapsack sprayer, users may find the step by step summary of the calibration process that follows easier to use. Permission is granted by the Forestry Commission to copy the following guide for personal use.

Calibration of a knapsack sprayer with a pressure control valve (PCV)

A Choose herbicide – refer to Field Book 8 and product labels

Note: Check whether use of knapsack sprayer is permissible and at what volume rate

Consider using the Forestry spot gun in preference to a knapsack sprayer, for spot applications

B Check required protective clothing – Field Book 8, Section 10.4, Table 10

C Measure spot diameter, from a test application using water with the lance held at a comfortable height. (1) [ ]

D Select volume rate (from the product label) (2) [ ] l/ha

E Use the following table to determine required volume per spot

Volume/spot

Spot diameter (metres)	Volume rate (litres/ha) (2)														
(1)	50	75	100	125	150	175	200	225	250	275	300	325	350	375	400
0.8					7.5	8.8	10.1	11.3	12.6	13.8	15.1	16.3	17.6	18.9	20.1
0.9				8.0	9.5	11.1	12.7	14.3	15.9	17.5	19.0	20.7	22.2	23.9	25.5
1.0			7.9	9.8	11.8	13.8	15.8	17.7	19.6	21.6	23.6	25.5	27.5	29.5	31.4
1.1		7.1	9.5	11.9	14.3	16.6	19.0	21.4	23.8	26.1	28.5	30.9	33.3	35.6	38.0
1.2		8.5	11.3	14.1	17.0	19.8	22.6	25.4	28.3	31.1	33.9	36.7	39.6		
1.3	6.7	10.0	13.3	16.6	19.9	23.3	26.5	29.9	33.2	36.5	39.8				
1.4	7.6	11.5	15.4	19.2	23.1	26.6	30.8	34.7	38.4						
1.5	8.8	13.3	17.7	22.1	26.5	30.9	35.4	39.8							

(3) [ ] ml

F Select nozzle

EITHER

(i) *For sprayers with a pressure control valve*

a) Time yourself spraying 10 spots (4i)

[ ] seconds

b) Multiply required volume per spot (3) by 10

i.e. (3) (5i)

[ ] × 10 = [ ] ml

c) Adjust knapsack pressure to deliver the volume required for 10 spots (5i) in the time taken to spray 10 spots (4i)

(ii) *For sprayers with no pressure control valve*

a) Measure nozzle output from a test application using water (4ii)

[ ] ml/min

b) Calculate required application time per spot

60 × [Required volume per spot (3) ÷ nozzle output (4ii)] = application time

i.e. (3) (4ii) (5ii)

60 × [[ ] ÷ [ ]] = [ ] seconds

G Practise application into a graduated measure until the desired volume per spot is achieved. Then spray over a dry surface to achieve the required spot size. If the application time is impractical, increase the volume rate and re-calibrate. A larger volume rate or different nozzle or reduced spray pressure may also be deemed more suitable depending on weather conditions, to reduce spray drift. In each case, re-calibration is required.

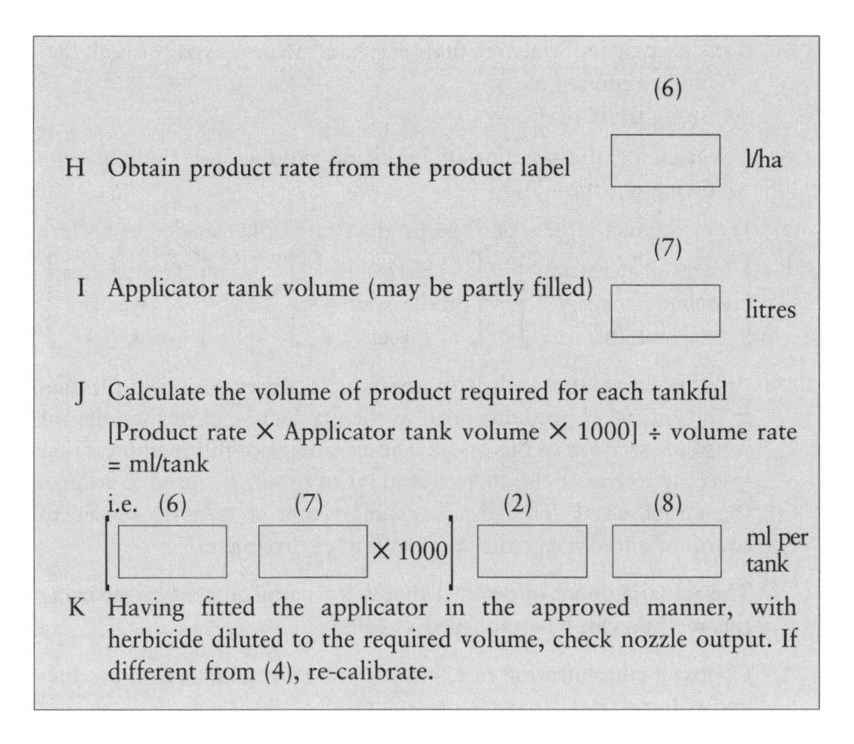

H	Obtain product rate from the product label	(6) [ ]	l/ha
I	Applicator tank volume (may be partly filled)	(7) [ ]	litres

J Calculate the volume of product required for each tankful
[Product rate × Applicator tank volume × 1000] ÷ volume rate = ml/tank

i.e. [(6) [ ] (7) [ ] × 1000] (2) [ ] (8) [ ] ml per tank

K Having fitted the applicator in the approved manner, with herbicide diluted to the required volume, check nozzle output. If different from (4), re-calibrate.

Application to individual bushes

When calibrating for overall, band or spot applications, weeds less than about 50 cm tall are assumed to have a negligible height – flat area is taken to equal surface area. However, when making applications to tall weeds or clumps of bushes – actual surface area must be considered. It is very difficult to calculate the surface area of bushes or tall weeds, so the following method is suggested to allow the calculation of reasonably accurate calibration rates.

1. Fill sprayer with water.
2. Mark out a representative area of clumps and bushes so that the foliage to be treated covers approximately 50 m^2.
3. Spray the area with water to wetness (the point just before run-off occurs).
4. Measure the exact amount of water required to refill the sprayer.
5. As the tank capacity is known the area that one full tank will cover can be calculated using the following formula:

$$\frac{\text{Tank capacity} \times \text{Trial area (ha)}}{\text{Volume required to spray trial area}} = \text{Treated area sprayed per tank (ha)}$$

6. To calculate the amount of herbicide product per tank use the following equation.

a. From product label – find the product rate of herbicide per hectare

b. $$\left[\begin{array}{c}\text{Litres of herbicide}\\ \text{required per sprayer-}\\ \text{tank full}\end{array}\right] = \left[\begin{array}{c}\text{Herbicide}\\ \text{product rate per}\\ \text{hectare}\end{array}\right] \times \left[\begin{array}{c}\text{Treated area}\\ \text{sprayed per}\\ \text{tank (ha)}\end{array}\right]$$

An alternative approach is to apply a % concentration of product in diluent, when recommended in the product label or the relevant herbicide sections of this book. The advantage of this method is that it is easy to ensure the correct amount of product is used to control the target weed. The disadvantage is that it is much easier to overdose and damage the surrounding environment.

The actual amount of product that will be applied per hectare using this method can be established as follows:

1. Choose a concentration rate, based upon label guidance on product and volume rates

$$\%\ \text{concentration} = \frac{\text{Product rate}}{\text{Volume rate}}$$

2. $$\text{Litres of herbicide required per sprayer-tank full} = \frac{\text{Product rate}}{\text{Volume rate}} \times \text{Sprayer tank capacity}$$

3. Make a test application, using water, as in steps 1–5 above.

4. $$\text{Actual product rate per hectare (l/ha)} = \frac{\text{Litres of herbicide required per sprayer-tank full}}{\text{Treated area sprayed per tank (ha)}}$$

Actual product rate applied must not exceed the maximum rate listed on the product label.

11.5.2 Dribble bar

Uses

For applying herbicides around sensitive tree species.

Supplier

Technical Engineering Ltd.
Unit 2
Henry Crabb Road
Littleport
Ely
Cambridgeshire
CB6 1SE
Telephone: 01353 862044

Description

The 0.5 m long bar has holes drilled at 2.5 cm spacing and it is fitted with a flow restrictor and a filter. It is manufactured to fit Cooper Pegler knapsack lances but adaptors for other lances are available. As the dribble bar is designed to work at low pressures it is essential that the knapsack is fitted with a pressure control valve.

Calibration for band and spot applications

Fit a dribble bar to the lance and fully pressurise the knapsack sprayer part filled with water. Reduce the pressure to the point where all of the holes are producing a constant stream of liquid and not breaking down into droplets. This should be in the region of 0.2 – 0.3 bar pressure (3–4 p.s.i.). It is essential during calibration and use that the dribble bar is kept level.

Band application

a. Measure dribble bar flow (ml) for one minute.

b. Determine the walking speed which an operator can maintain over the site with full applicator and wearing the necessary protective clothing.

c. Determine the volume rate from Table 15. Interpolate if necessary.

Table 15 Volume rates for the dribble bar applicator

Flow rate (ml/min)	600	650	700	750	800
Walking speed (m/min)	Volume rate (litres per hectare)				
60	200	217	233	250	267
55	218	236	255	273	291
50	240	260	280	300	320
45	267	289	311	333	356
40	300	325	350	375	400
35	343	371	400	429	457
30	400	433	467	500	533
25	480	520	560	600	640
20	600	650	700	750	800

d. Calculate quantity of product (herbicide) required per knapsack.

$$\text{Quantity of product required per knapsack (ml)} = \frac{\text{Knapsack capacity (l)} \times \text{Product rate (l/ha)} \times 1000}{\text{Volume rate (l/ha)}}$$

Example:

Dribble bar flow rate 700 ml/min
Walking speed 50 m/min
Knapsack capacity 15 litres
Product rate Roundup 2 l/ha

Therefore quantity of product required per knapsack (ml) $= \frac{15 \times 2 \times 1000}{280}$

$= 107$ ml

Spot application

a. Measure dribble bar flow for one minute.

b. Determine time to treat a 1 metre diameter spot ($0.79\ m^2$).

c. Calculate volume rate:

$$\text{Volume rate (l/ha)} = \frac{\text{Flow rate (ml/min)} \times \text{Time to treat spot (seconds)}}{\text{Spot area } (m^2) \times 6}$$

d. Calculate quantity of product (herbicide) required per knapsack:

$$\text{Quantity of product required per knapsack (ml)} = \frac{\text{Knapsack (l)} \times \text{Product rate (l/ha)} \times 1000}{\text{Volume rate (l/ha)}}$$

Example:

Flow rate	650 ml/min
Time to treat spot	4 seconds
Knapsack capacity	15 litres
Product rate	atrazine – 10 l/ha

Therefore volume rate (l/ha) $= \frac{650 \times 4}{0.79 \times 6}$

$= 549$ l/ha

Quantity of product required per knapsack (ml) $= \frac{15 \times 10 \times 1000}{549}$

$= 273$ ml

11.5.3 Fox Motori Electra F110 TS knapsack sprayer

Uses

Cut stump application at high volumes.
General herbicide application at medium and low volumes.

Supplier

Microcide Limited
Shepherds Grove
Stanton
Suffolk
IP31 2AR
Telephone: 01359 51077

Description

The Fox Motori Electra F110 TS is an electrically powered knapsack of 10 litres capacity, weighing 3.5 kg when empty. Fitted into the back of the knapsack are an electric pump, rechargeable, 6 volt battery and socket for battery recharger. The battery, under normal use, should last an 8 hour working day.

Tools required for maintenance and calibration

Pliers
Medium screwdriver

Adjustable spanner
Roll of PTFE plumbing tape
Plastic bucket
Funnel
Metric graduated measure
Absorbent paper
Watch
Tape measure
Spare nozzles

Calibration

The electric pump flow rate is non-adjustable. The only way to vary volume is to fit different size nozzle tips.

Table 16 gives an indication of the likely output of various nozzle types for different time control settings on the knapsack. It is important to measure the exact nozzle output to calculate the correct volume rate in litres/hectare.

For band and overall sprays use calibration calculations as for CP15 sprayer.

Spot spraying

There is insufficient pressure to give a metre diameter spot with a conventional solid cone nozzle. To give spot treatment with the Electra a floodjet/anvil nozzle giving a curtain of spray must be used to progressively treat the required spot. The application technique is: judge 0.5 m before the tree, press the lance switch and spray until 0.5 m beyond. The time control setting required to spray the metre long spot being determined by trial on a hard surface prior to actual pesticide application. To calibrate for spot application using the Lurmark AN floodjet/anvil nozzle:

i. Determine time control setting by undertaken trials to spray the required size spot.

ii. Select under the time control setting columns of Table 16 the required dose per spot to convert ml per dose to volume rate to comply with product label recommendations. To calculate volume rate in litres per hectare use the following equation:

$$\text{Litres per hectare} = \frac{10}{\text{Spot area (m}^2\text{)}} \times \text{dose (ml)}$$

Table 16

Nozzle output for Fox Motori Electra F110 TS knapsack sprayer

	Time control setting	1	2	3	4	5	6	7	8/9
Pesticide	**Spray time seconds**	**0.5**	**1.0**	**1.5**	**.20**	**2.5**	**3.0**	**3.5**	**Continuous**
	Nozzle flow				**ml/dose**				**ml/min**
	Nozzle tip								
	Lurmark								
Herbicide	AN 0.5	2.0	5.7	7.9	9.7	12.0	16.0	17.9	340
(weeding band,	AN 0.75	2.2	7.3	9.8	12.7	16.5	20.0	23.3	480
overall & spot	AN 1.0	2.8	8.4	11.7	14.7	18.2	24.4	27.2	520
with floodjet/	AN1.5	3.3	9.9	13.4	17.3	20.7	27.4	31.7	550
anvil nozzle)	AN 2.0	4.0	11.0	15.0	19.1	21.7	30.0	34.0	610
	AN 2.5	4.1	11.5	15.4	19.5	23.3	32.5	35.5	640
	Dribble bar	–	–	–	–	–	–	–	770
Herbicide (woody weeds spot application)	Spraying system TG2 (solid cone)	4.3	10.4	14.6	18.0	23.0	28.7	32.1	590
Insecticide (spot application top-up spray)	Spraying system 0002 (solid stream)	2.1	8.1	12.1	15.1	18.2	24.2	27.5	500
	Spraying system TG1 (solid cone)	3.0	8.1	11.4	14.0	17.4	23.0	25.7	480

11

Example:

Spot area 1 m^2

Dose Time control setting 3. AN 1.5 nozzle = 13.4 ml/dose.

$$\text{Litres per hectare} = \frac{10}{1} \times 13.4 = 134 \text{ l/ha}$$

iii. Measure actual nozzle flow and if different from Table 14 recalculate volume rate.

iv. Calculate the product quantity needed per knapsack fill.
e.g. glyphosate at 2 l/ha

$$\text{Requirement per fill} = \frac{10 \text{ (knapsack capacity)} \times 2 \times 1000}{134} = 149 \text{ ml}$$

11.5.4 Tecnoma T18P knapsack sprayer

Use

Ammonium sulphamate application to cut stump at high volume and following frill girdling.

Supplier

Technoma Agricentre
Mill Lane
Great Massingham
Norfolk
PE32 2HT
Telephone: 01485 520686

Description

Tecnoma T18P; 18 litre knapsack sprayer filled with a stainless steel ball bearing in the pump which will not be corroded by ammonium sulphamate. The return spring in the trigger assembly will corrode and will require replacement.

Tools required for maintenance

No tools are required as the sprayer can be dismantled by hand. Plastic bucket and absorbent paper should be available.

Calibration

As the sprayer has no pressure controls and as ammonium sulphamate is applied to the point of run-off, no calibration is necessary.

Cleaning

Spray out all dilute pesticide safely (see Section 3.9).

Wash and flush through with a weak solution of warm detergent and rinse with clean water several times.

11.5.5 Forestry spot gun

Uses

Cut stump application at high volume.

Woody weed foliar application at low volume.

Grasses and herbaceous broadleaved application at low and very low volume.

Supplier

Selectokil
Abbey Gate Place
Tovil
Maidstone
Kent
ME15 0PP
Telephone: 01622 755471

Description

Forestry spot gun; adapted from the veterinary drench gun with a 5 litre knapsack. This applicator can be adjusted to give a measured dose up to 20 ml.

Accessories

Spare knapsack from Selectokil.

Steel cylinder for use with certain pesticides from Selectokil.

Spare nozzle from Selectokil.

25 litre rigid polythene bottle with 18 mm tap bore.

Ref. BA/PS4

From:

Fisons Scientific Equipment Division
Bishops Meadow Road
Loughborough
Leicestershire
LE11 0RG

Nozzle data

Nozzle type	For treating
TG 2.8W wide angle solid cone	Grass/herbaceous broadleaved weed
TG5 narrow angle solid cone	Woody weeds
No. 0006 or LF6-0 solid stream	Chemical thinning or cut stump

Tools required for maintenance and calibration

Pliers
Two 25 mm adjustable wrenches
Funnel
Plastic bucket
Metric graduated measure
Absorbent paper
Tape measure
Lubricating oil
Roll of PTFE plumber's tape

Calibration

1. Adjust gun to deliver a 5 ml dose.
2. Fill knapsack with 5 litres of water and mark this level (filler cap upmost).
3. Prime gun.
4. With the nozzle held at spray height over a dry surface, squeeze the trigger a number of times until a well defined spot appears. Measure spot diameter, disregarding peripheral droplets.
5. Enter Table 17 by spot diameter and application rate of herbicide product to determine the quantity of product required per 5 litre knapsack.

Example

Spot diameter 1.2 m.
Rate of application for glyphosate to control grasses in lowland Britain is 1.5 litres per treated hectare.

Therefore from Table 17 the quantity of herbicide required per 5 litres is 0.17 litres (ie 0.17 litres of herbicide + 4.83 litres of water = 5 litres).

6. Fill knapsack in the approved manner with herbicide/diluent mixture.
7. Check the spot diameter as in paragraph 4; if different re-calibrate.

Note:
When doses larger than 5 ml are necessary, it is suggested that several small doses up to the required volume are applied to minimise fatigue to the hands. The amount of herbicide product added to each 5 litre knapsack will need to be altered. This can be calculated using the following equation.

Required quantity of herbicide product =

$$\frac{\text{Quantity of product for 5 ml dose}}{5} \times \text{new dose rate per 5 litre knapsack}$$

Cleaning

Spray out all dilute pesticide safely (see Section 3.9).
Wash and flush through with a weak solution of warm water and detergent, rinse with clean water.

Note: damage to the applicator can occur if subject to temperature in excess of 50°C.

To lubricate, split the gun by undoing the handle cramp screw, remove the complete piston assembly and apply a liberal quantity of oil to the moving parts.

11.5.6 Microfit Herbi

Use

Band or overall application at very low volumes.

Supplier

CDA Ltd.
Three Mills
Bromyard
Herefordshire
HR7 4HU
Telephone: 01885 482397

Description

Microfit Herbi is a low speed rotary atomiser producing droplets with a volume median diameter (VMD) of 250 microns.

Accessories

2.5 litre plastic bottles from CDA Ltd.
Spare motor or atomiser heads from CDA Ltd.

Table 17

Quantity of herbicide product required per 5 litre knapsack – forestry spot gun

Spot diameter (metres)	Product rate (herbicide product litres per hectare)																	
	1.0	**1.5**	**2.0**	**3.0**	**3.75**	**4.0**	**5.0**	**6.0**	**7.0**	**8.0**	**9.0**	**10.0**	**11.0**	**12.0**	**13.0**	**13.5**	**14.0**	**15.0**
0.60	0.03	0.04	0.06	0.08	0.11	0.11	0.14	0.17	0.20	0.23	0.25	0.28	0.31	0.34	0.37	0.38	0.40	0.42
0.65	0.03	0.05	0.07	0.10	0.12	0.13	0.17	0.20	0.23	0.27	0.30	0.33	0.37	0.40	0.43	0.45	0.46	0.50
0.70	0.04	0.06	0.08	0.12	0.14	0.15	0.19	0.23	0.27	0.31	0.35	0.38	0.42	0.46	0.50	0.52	0.54	0.58
0.75	0.04	0.07	0.09	0.13	0.17	0.18	0.22	0.27	0.31	0.35	0.40	0.44	0.49	0.53	0.57	0.60	0.62	0.66
0.80	0.05	0.08	0.10	0.15	0.19	0.20	0.25	0.30	0.35	0.40	0.45	0.50	0.55	0.60	0.65	0.68	0.70	0.75
0.85	0.06	0.09	0.11	0.17	0.21	0.23	0.28	0.34	0.40	0.45	0.51	0.57	0.62	0.68	0.74	0.77	0.79	0.85
0.90	0.06	0.10	0.13	0.19	0.24	0.25	0.32	0.38	0.45	0.51	0.57	0.64	0.70	0.76	0.83	0.86	0.89	0.95
0.95	0.07	0.11	0.14	0.21	0.27	0.28	0.35	0.43	0.50	0.57	0.64	0.71	0.78	0.85	0.92	0.96	0.99	1.06
1.00	0.08	0.12	0.16	0.24	0.29	0.31	0.39	0.47	0.55	0.63	0.71	0.79	0.86	0.94	1.02	1.06	1.10	1.18
1.05	0.09	0.13	0.17	0.26	0.32	0.35	0.43	0.52	0.61	0.69	0.78	0.87	0.95	1.04	1.13	1.17	1.21	1.30
1.10	0.10	0.14	0.19	0.29	0.36	0.38	0.48	0.57	0.67	0.76	0.86	0.95	1.05	1.14	1.24	1.28	1.33	1.43
1.15	1.10	0.16	0.21	0.31	0.39	0.42	0.52	0.62	0.73	0.83	0.93	1.04	1.14	1.25	1.35	1.40	1.45	1.56
1.20	0.11	0.17	0.23	0.34	0.42	0.45	0.57	0.68	0.79	0.90	1.02	1.13	1.24	1.36	1.47	1.53	1.58	1.70
1.25	0.12	0.18	0.25	0.37	0.46	0.49	0.61	0.74	0.86	0.98	1.10	1.23	1.35	1.47	1.60	1.66	1.72	1.84
1.30	0.13	0.20	0.27	0.40	0.50	0.,53	0.66	0.80	0.93	1.06	1.19	1.33	1.46	1.59	1.73	1.79	1.86	1.99
1.35	0.14	0.21	0.29	0.43	0.54	0.57	0.72	0.86	1.00	1.15	1.29	1.43	1.57	1.72	1.86	1.93	2.00	2.15
1.40	0.15	0.23	0.,31	0.46	0.58	0.62	0.77	0.92	1.08	1.23	1.39	1.54	1.69	1.85	2.00	2.08	2.16	2.31
1.45	0.17	0.25	0.33	0.50	0.62	0.66	0.83	0.99	1.16	1.32	1.49	1.65	1.82	1.98	2.15	2.23	2.31	2.48
1.50	0.18	0.27	0.35	0.53	0.66	0.71	0.88	1.06	1.24	1.41	1.59	1.77	1.94	2.12	2.30	2.39	2.47	2.65
1.55	0.19	0.28	0.38	0.57	0.71	0.75	0.94	1.13	1.32	1.51	1.70	1.89	2.08	2.26	2.45	2.55	2.64	2.83
1.60	0.20	0.30	0.40	0.60	0.75	0.80	1.00	1.21	1.41	1.61	1.81	2.01	2.21	2.41	2.61	2.71	2.81	3.02

Battery tester with voltage ranges of 0–1.5 v and 0–12 v.
Panda Multimeter Model ELT 801 (obtain from a local electrical retailer).
Alkaline manganese batteries (obtain locally).

Tools required for maintenance and calibration

Pliers
Philips screwdriver
Plastic bucket
Large funnel
Metric graduated measure
Absorbent paper
Watch
50 m tape measure

Calibration

1. Determine the walking speed which an operator can maintain over the site with a full applicator and wearing the necessary protective clothing.
2. Select volume rate from the Method of application paragraph of the selected herbicide.
3. Enter Table 18 by walking speed and volume rate to find the required nozzle output (ml/min) over a 1.2 metre swathe.

   The Micron Herbi-4 is a new addition to the Micron range that can be used with larger 5 or 10 litre knapsacks. Use and calibration is very similar to those outlined for the Herbi below, although nozzle outputs may be slightly higher due to increased liquid pressure. Refer to Technical Development Branch Information Note 8/93 for further details.
4. The ideal feed rate to the atomisers is 60 ml/min. At 120 ml/min and above, atomisation is not so efficient; small satellite droplets may form and present a drift hazard.
5. *Always check swathe width*
   For a swathe wider than 1.2 metres the calculated nozzle output (from Table 18) has to be adjusted by the following equation:

$$\text{Required nozzle output for new swathe width} = \frac{\text{Required nozzle output for 1.2 m swathe (ml/min)} \times \text{Actual swathe width (m)}}{1.2}$$

Table 18 Required nozzle output for 1.2 metre swathe (ml/min) – Microfit Herbi

Sustainable walking speed for site (ml/min)	Volume rate (l/ha)											
	10.0	12.0	14.0	16.0	18.0	20.0	22.0	24.0	26.0	28.0	30.0	32.0
60	72	87	101	116								
55	66	79	93	106	119							
50	60	72	84	96	108	120						
45		65	76	85	98	108	119					
40		58	67	77	87	96	106	116				
35			59	67	76	84	93	101	110	118		
30				58	65	72	80	87	94	101	108	116
25						60	66	72	78	84	90	96
20								58	63	68	72	77
15												58

Note: Interpolate where necessary.

Example:

Walking speed	40 m/min
Volume rate	20 l/ha
Actual swathe width	1.4 m

From Table 18 the required nozzle output for a 1.2 metre swathe is 96 ml/min. Therefore the required nozzle output for a 1.4 metre swathe width is:

$$\frac{96 \times 1.4}{1.2} = 112 \text{ ml/min}$$

6. Remove the spinning disk from the applicator
7. Fit the feed nozzle which has the nearest nozzle output to that calculated.

Nozzle colour	Nominal nozzle output with water
Blue	90 ml/min
Yellow	150 ml/min
Orange	240 ml/min
Red	360 ml/min

Note:

Viscosity of liquid will alter flow rates.

8. Attach a bottle containing the mixture of herbicide to be sprayed to the applicator.
9. Hold the applicator at an angle of about 40 degrees. Before placing the nozzle over a graduated measure ensure that all air bubbles are removed from the nozzle in order to obtain a constant flow. Measure the nozzle output for a set time to obtain nozzle output in ml/min.
10. Check that the measured output agrees with that calculated in paragraph 3 or 5. If they do not agree re-calibrate by either:

a. if the outputs are close, adjust the angle of the applicator. Increasing the angle up to 10 degrees will increase the flow and reducing the angle up to 10 degrees will reduce the flow. If this fails to give the required output, fit a different nozzle of the same colour.

Note:

There can be considerable variation in nozzle output within the same colour nozzles.

or b. if there is a considerable difference in output, fit a nozzle of different colour;

or c. adjust volume rate if (a) and (b) do not achieve the required nozzle output;

or d. if (b) and (c) fail to achieve the required nozzle output adjust walking speed.

11. Check swathe width and re-calibrate if necessary.
12. Refit spinning disc.

Droplet size

Approximately 250 micron droplet VMD with the spinning disc revolving at 2000 rpm providing the voltage is between 4 and 6 volts. Check the voltage at the plug between the battery carrier and lance assembly; if below 4 volts replace batteries.

Herbi 77 is no longer manufactured but many are still in use; treat as Microfit Herbi except for voltage range. The voltage range for this applicator is 4 to 12 volts. Check the voltage at the two electrical terminals on the outside of the atomiser head; if below 4 volts replace batteries.

Cleaning

Spray out all dilute pesticide safely (see Section 3.9).
Remove batteries and clean off any corrosion on the contact points before storage. Wash thoroughly with a weak solution of detergent and rinse with clean water.

11.5.7 ULVA 8/ULVA +

Use

Incremental application.

Supplier

CDA Ltd.
Three Mills
Bromyard
Herefordshire
HR7 4HU
Telephone: 01885 482397

Description

High speed (5500–8800 rpm) rotary atomiser producing droplets with a volume median diameter (VMD) of 70 microns.

ULVA 8 powered by 8 batteries. Its use is for incremental spraying in crops up to an average eight of 1.5 m and for inter-row spraying of heather.

The ULVA + has been recently added to the Micron range to replace the ULVA 8. The unit is powered by six HP2 batteries. Calibration is carried out in the same way as the ULVA 8 but nozzle flow rates are different.

Accessories

Spare 1 litre plastic bottles from CDA Ltd.
Vibrotac high range rev counter (one per gang) from CDA Ltd.
Spare motor, spinning disc or complete head assembly from CDA Ltd.
Batteries: alkaline manganese batteries are recommended. If these batteries are used for 4 hours, then rested for about 20 hours before being used again they will give 25–30 hours total spraying time. Purchase locally. For larger programmes, rechargeable cadmium nickel batteries can be used.

Tools required for maintenance and calibration

Pliers
Screwdriver 2½" and 3"
Screwdriver No. 2 Pozi 5332
Small adjustable wrench
Roll of adhesive tape
Plastic bucket
Large funnel
Metric graduated measure
Absorbent paper
Watch
50 m tape measure

Calibration

Incremental application

1. Determine the walking speed which an operator can sustain over the site with a full applicator and wearing the necessary protective clothing.
2. Select volume rate from the Method of application paragraph of the selected herbicide.
3. Enter Table 19 by walking speed and volume rate to find the required nozzle output (ml/min) over a 5 metre swathe. If the combination of volume rate and walking speed results in a required nozzle output outside the range actually possible for available nozzles, select an alternative volume rate.
4. Although the ideal nozzle output is 60 ml/min there are some circumstances where, in order to avoid walking at an extremely slow

speed, the flow rate can be allowed to rise to 120 ml/min or even 180 ml/min, even though this will result in larger, rather than more, droplets being formed.

Table 19 Nozzle output for ULVA 8 (ml/min)

Sustainable walking speed for site (m/min)	Volume rate (l/ha)					
	10.0	12.0	14.0	16.0	18.0	20.0
36	180					
34	170					
32	160					
30	150	180				
28	140	168				
26	130	156	182			
24	120	144	168			
22	110	132	154	176		
20	100	120	140	160	180	
18	90	108	126	144	162	180
16	80	96	112	128	144	160
14	70	84	98	112	126	140
12	60	72	84	96	108	120
10	50	60	70	90	90	100

5. If because of distance between rows a different swathe width is required within the operating range of 4–6 metres, the required nozzle output (from Table 19) has to be adjusted by the following equation:

$$\text{Required nozzle output for new swathe width (ml/min)} = \text{Required nozzle output for 5 m swathe (ml/min)} \times \text{Actual swathe width(m)} \times 0.2$$

Example:

Walking speed	26 m/min
Volume rate	12 l/ha
Actual swathe width	4.7 m

From Table 19 the required nozzle output for a 5 metre swathe is 156 ml/min.

Therefore the required nozzle output for a 4.7 metre swathe width is:

$$156 \times 4.7 \times 0.2 = 147 \text{ ml/min.}$$

6. Remove the spinning disc from Ulva.
7. Fit the feed nozzle which has the nearest nozzle output to that calculated.

ULVA 8

Nozzle colour	Nominal nozzle output with water
Yellow	30 ml/min
Red	60 ml/min
Grey	120 ml/min
Green	180 ml/min

Note:

Viscosity of liquid will alter flow rates.

ULVA +

Nozzle colour	Nozzle output using Asulox/Actipron
Red	60 ml/min
Black	95 ml/min
Grey	115 ml/min
Pink	135 ml/min

8. Attach a bottle containing the mixture of the herbicide to be sprayed to the applicator. (Use equation 10 in Section 11.3.3 to calculate the quantity of herbicide required per applicator container.)
9. Before placing the nozzle over a graduated measure ensure that all air bubbles are removed from the nozzle in order to obtain a constant flow. Measure the nozzle output for a set time to obtain nozzle output in ml/min.
10. Check that the measured output agrees with that calculated in paragraphs 3 or 5. If they do not agree re-calibrate by either:

a. if the measured and required outputs are close, fit a different nozzle of the same colour;

Note:

There can be considerable variation in nozzle output within the same colour nozzles.

or b. if there is a considerable difference in output fit a nozzle of a different colour;

11

or c. adjust volume rate if (a) and (b) do not achieve the required nozzle output;

or d. if (b) and (c) fail to achieve the required nozzle output adjust walking speed.

11. If during the day the temperature changes, the nozzle output must be checked and if necessary re-calibrated.

Directed application – heather control

1. Determine the walking speed which an operator, whilst sweeping the applicator from side to side, can sustain over the site with a full applicator and wearing the necessary protective clothing.
2. Select volume rate from the Method of application paragraph of the selected herbicide.
3. Enter Table 20 by walking speed and volume rate to find required nozzle output (ml/min) over a 2 metre swathe.
4. Although the ideal nozzle output is 60 ml/min there are some circumstances where, in order to avoid walking at an extremely slow speed, the flow rate can be allowed to rise to 120 ml/min or

Table 20 Required nozzle output for 2 metre swathe (ml/min) ULVA 8

Sustainable walking speed for site (m/min)	Volume rate (l/ha)			
	10.0	12.0	15.0	16.0
40	80	96	120	128
38	76	91	114	122
36	72	87	108	116
34	68	82	102	109
32	64	77	96	103
30	60	72	90	96
28	56	68	84	89
26	52	63	78	84
24	48	58	72	77
22	44	53	66	71
20	40	48	60	64
18	36	44	54	58
16	32	39	48	52
14	28	34	42	45
12	24	29	36	39
10	20	24	30	32

even 180 ml/min, even though this will result in larger, rather than more, droplets being formed.

5. If a different swathe width is required the nozzle output (from Table 20) has to be adjusted by the following equation:

$$\text{Required nozzle output for new swathe width} = \frac{\text{Required nozzle output for 2 m swathe (ml/min)} \times \text{Actual swathe width (m)}}{2}$$

Example:

Walking speed 18 m/min
Volume rate 12 l/ha
Actual swathe width 1.75 m

From the table the required nozzle output for a 2 metre swathe is 44 ml/min. Therefore the required nozzle output for a 1.75 metre swathe width is:

$$\frac{44 \times 1.75}{2} = 38.5 \text{ ml/min}$$

6. Remove the spinning disc from Ulva.
7. Fit the feed nozzle which has the nearest nozzle output to that calculated.

Nozzle colour	Nominal nozzle output with water
Yellow	30 ml/min
Red	60 ml/min
Grey	120 ml/min
Green	180 ml/min

Note:
Viscosity of liquid will alter flow rates.

8. Attach a bottle containing the mixture of the herbicide to be sprayed to the application. (Use equation 10 in section 11.3.3 to calculate the quantity of herbicide required per applicator container).
9. Before placing the nozzle over a graduated measure ensure that all air bubbles are removed from the nozzle in order to obtain a constant flow. Measure the nozzle output for a set time to obtain nozzle output in ml/min.
10. Check that the measured output agrees with that calculated in paragraph 3 or 5. If they do not agree re-calibrate by either:

a. if the measured and required output are close, fit a different nozzle of the same colour.

Note:
There can be considerable differences in nozzle output within the same colour nozzles.

or b. if there is a considerable difference in output, fit a nozzle of a different colour;

or c. adjust volume rate if (a) and (b) do not achieve the required nozzle output;

or d. if (b) and (c) fail to achieve the required nozzle output adjust walking speed.

11. Refit spinning disc.

12. If during the day the temperature changes, the nozzle output must be checked and if necessary re-calibrated.

Droplet size

The droplet size is approximately 70 microns VMD when the disc speed is between 5500–8000 rpm. Disc speed should be checked regularly with the 'Vibrotac'; if below 5500 RPM the batteries should be changed.

Wind speed

Wind speed should always be measured at the height of the sprayer.

For incremental application over an approximate 5 metre swathe, a minimal speed of 3 km per hour (2 mph) is acceptable with a maximum speed of 12 km per hour (7 mph), occasionally gusting to 16 km per hour (10 mph).

For directed application on heather 0–19 km per hour (0–12 mph) is suitable.

Wind direction

Incremental application can be carried out if the angle between the operator's line of walking and the wind direction is not less than 20 degrees.

Safety zone

A downwind safety zone of at least 100 metres, should be allowed where there are susceptible crops, especially those on adjoining farmland and crops growing in polytunnels or under glass. Do not spray if the wind is blowing towards susceptible glass house and similar crops.

Cleaning

Remove batteries and clean off any corrosion on the contact points before storage.

Wash thoroughly with a weak solution of detergent and rinse with clean water.

11.5.8 Handheld direct applicator Weedwiper Mini

Supplier

Horitchem Limited
14 Edison Road
Churchfield Industrial Estate
Salisbury
Wiltshire
Telephone: 01722 320133

Description

SM90 Weedwiper Mini Standard Single-head (Standard Blue Label Wick) fitted with venting cap.

SMR90 Weedwiper Mini Red Band Single-head (Fast Flow Red Label Wick) fitted with venting cap.

The SM90 fitted with the standard wick is recommended for use on normal vegetation. Where the vegetation is dense or semi-resistant, the SMR90 with the fast flow wick should be used.

Accessories

Spare wicks from: Hortichem Limited.
10 × 55 g sachets of red dye from Hortichem Limited.
5-litre (Ref. PO 23/01) or 10 litre (Ref. PO 23/02) graduated polythene bottle with screw cap, tap and metal carrying handle.
From:
Fisons Scientific Equipment Division
Bishop Meadow Road
Loughborough
Leicestershire
LE11 0RG

Tools required for maintenance and calibration

Small adjustable wrench
Small funnel to fit filler cap
Metric graduated measure
Absorbent paper
Plastic bucket

Teaspoon
Scrubbing brush
Plastic bags

Calibration

As it is not possible to calibrate the weedwiper, the only way of ensuring that the wick is wet enough to transfer herbicide on the vegetation is by adjusting the flow during use.

To adjust flow:

1. Mark the reservoir (handle) at 75 cm from the filler cap. Always keep the level of herbicide above this point during use to ensure adequate wick flow.
2. Fill applicator with herbicide mixture to which has been added 1 teaspoonful of red dye per litre.
3. Open venting cap.
4. Allow wick to become impregnated with herbicide mixture. When fully impregnated the wick should be red.
5. Start treating the vegetation (see Working method).
6. If the wick blanches during use, the compression nuts should be released to increase wick flow. It may also be necessary to pull the rubber seals from their seats and re-adjust their position on the wick.
7. If the wick is dripping excessively, tighten the compression nuts to reduce wick flow.

Working method

A back and forth sweeping action, taking care not to touch the crop tree, will ensure that both sides of the weeds are treated. Where the trees are standing above the weeds, application should be made in a triangular pattern. A triple width pass on each side of the triangle will give control over a 1.08 m diameter spot. For trees below weed level, use 1.0 m long, back and forth passes over the tree, ensuring there is a 10 cm clearance between the wick and the top of the tree, to give control over 0.90 square metre.

Cleaning

In order to achieve maximum transfer of herbicide on to the weeds, any dirt adhering to the wick should be scrubbed off immediately. It is recommended that each day after use the applicator head is either immersed in water or the applicator filled with water and left soaking overnight to cleanse the wick.

Wicks should be replaced regularly.

New wicks should be fitted after storage or if the old wicks have dried out.

Wash thoroughly with a weak solution of detergent and rinse the weedwiper with clean water.

11.5.9 Ulvaforest tractor-mounted sprayer

Use

Tractor-mounted sprayer for band or complete application at very low volume.

Supplier

Controlled Droplet Application Ltd.
Three Mills
Bromyard
Hereford
HR7 4HU
Telephone: 01235 833270

Note: These sprayers are only built to order.

Description

The Ulvaforest consists of a 270 litre tank and a fully gimballed 4.8 metre boom mounted on to a hydraulic ram which allows it to be raised to 2 metres. Five Microfit Herbi rotary atomisers are fitted to the boom, each being controlled by an individual switch in the tractor cab. There are three basic spraying patterns, i.e. overall with a 6 metre swathe or over-the-rows spraying of either two or three rows with a nominal 1.2 metre swathe.

Maintenance and calibration

Details can be obtained from:
Technical Development Branch
Ae Village
Dumfries
DG1 1QB
Telephone: 01387 860264

11.6 Applicators for spot application of granules

11.6.1 Pepperpots

These are supplied free with propyzamide granules.

Accessories

A belt to hold 13 propyzamide beaker-type pots can be obtained from:
Kingswood Canvas Ltd.
197 Two Mile Hill Road
Kingswood
Bristol
BS15 1AZ

For transportation the lids of the pepperpots should be sealed with adhesive tape. Exchange the sealed lid with an unsealed one for application.

Calibration

1. Rate of application.
   Application rate
   To calculate g/m², divide rate in kg/ha by 10.
   e.g.

Per hectare		Per m²
30.0 kg/ha	=	3.0 g/m²
38.0 kg/ha	=	3.8 g/m²
40.0 kg/ha	=	4.0 g/m²

2. To calculate the quantity of granules required to treat a spot greater or lesser than one square metre, multiply the actual spot area to be treated by the quantity of product needed to treat 1 m².

   Example:
   Quantity of granules required to treat a 1.2 m² spot at 38.0 kg/ha is:

$$3.8\text{g} \times 1.2 \text{ m} = 4.56 \text{ g per spot}$$

Applicator calibration

1. Fill applicator with granules.
2. Using the side to side application technique shake the applicator for a set number of times over a paper or plastic sheet.

3. Use the graduated scale on the side of the pot to ensure application is at the specified rate.
4. Calculate whether the amount of granules applied is correct for the area treated (i.e. spot area × number of spots treated).

Cleaning

Wipe the inside and outside of the applicator with a dry paper towel.

11.6.2 Tree Mate Kerb granule applicator

Uses

Granular herbicide applicator.

Supplier

Pan Britannica Industries Ltd.
Britannica House
Waltham Cross
Herts
EN8 7DY
Telephone: 01992 23691

Description

The applicator comprises of a knapsack hopper, which will hold one 2000 tree bag of Kerb granules, a hand held metering unit and a flexible tube which connects the hopper to the metering unit. The metering unit has a 3.8 g chamber, operating button, on/off supply switch, restrictor delivery tube. The empty applicator weighs approximately 2.5 kg.

Tools required for maintenance and calibration

Screwdriver
Paper towels

Applicator calibration

This applicator delivers a measured dose per spot by releasing the contents of the 3.8 g chamber.

Applicator technique

Shake the delivery tube from side to side whilst traversing the spot.

11.7 Tree injection applicator

Forestry spot gun

Uses

Cut stump treatment and stem injection at high volume. Spot treatment of foliage at low volume.

Supplier

Selectokil
Abbey Gate Place
Tovil
Maidstone
Kent
ME15 0PP
Telephone: 01622 755471

Description

Forestry spot gun; adapted from the veterinary drench gun with a 5 litre knapsack. This applicator can be adjusted to give a measured dose up to 20 ml.

Accessories

Spare knapsack from Selectokil.
Steel cylinder for use with certain pesticides from Selectokil.
Spare nozzles from Selectokil.
25 litre rigid polythene bottle with 18 mm tap bore.
Ref. BA/PS4
From:
Fisons Scientific Equipment Division
Bishops Meadow Road
Loughborough
Leicestershire
LE11 0RG

Nozzle data

Nozzle type	For treating
TG 2.8 W TG5 No. 0006 or LF6–0 solid stream	Chemical thinning or cut stump

Calibration

For stem injection purposes a percentage solution of herbicide product is applied – see relevant herbicide sections.

Tools required for maintenance and calibration

Pliers
Two 25 mm adjustable wrenches
Funnel
Plastic bucket
Metric graduated measure
Absorbent paper
Lubricating oil
Roll of PTFE plumber's tape

11.8 ENSO brush-cutter stump treatment attachment

Use

Application of herbicide directly on to cut stump during cutting operation.

Supplier

Chieftan Forge
Burnside Road
Bathgate
West Lothian
EH48 4PH
Telephone: 01506 52354

When ordering, detail make and model number of brush cutter.

Description

The kit consists of a supply tank, pump operated by a twist grip, nozzle and pipe-work for fitting to the brush-cutter. In use, by twisting the grip and activating the pump, herbicide is sprayed out on to the underside of the cutting blade thus depositing herbicide on to the freshly cut stump whilst cutting.

Maintenance and calibration

Details of maintenance and calibration can be obtained from:
Forestry Commission
Technical Development Branch
Ae Village

Dumfries
DG1 1QB
Telephone: 01387 860264

11.9 Pesticide transit box

Use

Box suitable for transporting pesticides in vehicles.

Internal dimensions

Length	53 cm
Width	37 cm
Height	35 cm

Supplier

Cairn Craft
The Factory
Foyers (by Loch Ness)
Inverness
Telephone: 0145 63285

11.10 Output guides – herbicide application

This section contains two output guides which can be used to estimate the time taken to apply herbicides in forest situations. The guides will give a target time per hectare or a time in 'standard' minutes per hectare for herbicide application. These times take into account all aspects of the job, such as the job specification, rest allowance, etc. – it is essential that the guides are worked through in a systematic way from start to finish to allow an accurate estimate of standard/target time.

The output guides give a figure of how long it will take to treat 1 ha of land.

Costs can be estimated from time taken to treat 1 ha – simply multiply expected hourly labour costs by the time taken to treat a hectare.

$$\text{Cost/ha} = \underset{\text{(from market rates)}}{\text{Cost/hour}} \times \frac{\text{Time in minutes}}{60} \text{ (from output guide)}$$

Alternatively, the standard output per hour can be calculated as follows:

$$\text{Standard output/hour} = \frac{60}{\text{Time/ha}} \quad \text{(from output guide)}$$

Costs can be calculated from standard output in a similar way:

$$\text{Cost/ha} = \frac{\text{Cost/hour}}{\text{Standard output/hour}} \quad \text{(market rates)}$$

Example:

From output guide, time taken to treat 1 ha = 259 minutes.

Current market hourly rates are known to be £5/hour.

$$\text{Cost/ha} = 5 \times \frac{259}{60}$$

$$= £21.58/\text{ha}$$

$$\text{Alternatively, standard output/hour} = \frac{60}{259}$$

$$= 0.23 \text{ ha/hour}$$

$$\text{Cost/ha} = \frac{5}{0.23}$$

$$= £21.58/\text{ha}$$

These cost figures can be used in several ways.

1. To help cost a job and build up a budget (see Section 1.8 for herbicide costs).
2. To aid interpretation of tendered prices.
3. To assist in setting prices for piecework operations.

Forestry Commission staff should refer to Industrial Personnel Memorandum 25 for current policy on calculating piecework prices from standard times and output.

11.11 Output guide – overall or band application (Weeding 1, Jan 1990)

Herbicide application

Applications by knapsack – overall or band

ULVA – overall

Microfit Herbi/Herbi – overall or band

1. Conditions

These times apply to the application of herbicides under the following conditions:

a. Pre- and post-planting.

b. All terrain and ground conditions.

c. An adequate supply:
i. for knapsack, separate herbicides and diluent;
ii. for ULVA, Microfit Herbi/Herbi pre-bottled or pre-mixed herbicides.

The above should be distributed on site to avoid excessive walking to refill applicator (within 30 metres of spraying site).

d. Vegetation and crop height suitable for the particular applicator and application method.

e. Times are per plantation hectare.

f. 'Row' spacing between 1.0 m and 3.0 m (up to 6 m/a ULVA drift application).

g. Trees either visible or not visible.

2. Job Specification

The times are for the following work:

a. Applying the prescribed volume rate through the selected applicator.

b. General preparation, on/off protective clothing and equipment, wash down clothing, walk to and from rest camp, check calibration, maintain and wash out applicators.

c. Walk to and from the spraying site to the supply point to refill applicator (Para 7).

3. Tools and equipment

a. i. 15 litre knapsack;

or ii. ULVA;

or iii. Microfit Herbi/herbi.

b. Tools for maintenance and calibration as listed against the corresponding applicator in Section 11 of this book.

c. Equipment required for mixing and filling applicators:

i. For mixing herbicide solutions on site for the knapsack, a plastic bucket, funnel and metric graduated measure.

ii. For pre-mixed herbicide supplied in a suitable plastic container with tap, a funnel.

d. First Aid Kit as per FASTCo Safety Guide 34 (see Section 3.2) – each operator to have a personal Emergency Kit A in the immediate work area and for there also to be a squad Emergency Kit B available on work site.

e. Protective clothing and equipment in accordance with Section 10 of this book.

f. Washing facilities for personal hygiene and wash down of protective clothing and equipment.

4. *Allowances*

Allowances for other work and rest are given below. They cover non direct work such as general preparation, putting on and taking off protective clothing (including washing down), calibration of equipment; and allowance for personal needs and rest.

Application	Other work allowance (%)	Rest allowance (%)	Overall allowance factor
Knapsack	10	22	1.34
ULVA	11	18	1.31
Microfit Herbi/Herbi	10	16	1.28

5. *Method of using the output guide*

a. On site measure distance between rows (or swathe width during overall application).

b. On site, wearing the specified clothing and equipment and with an applicator full of herbicide or water, walk and apply herbicide (water) in the prescribed manner for one minute. Measure the distance covered in that minute. If conditions vary over the site repeat the timing exercise sufficiently to get an acceptable average of distance covered per minute. (*Note:* For the ULVA, Herbi, and Knapsack sprayers the walking speed achieved in calibration must be adhered to during application).

c. Use row spacing for entry to para 6a and divide the answer by the average distance covered per minute calculated as in para b. above. The result is basic time per hectare. For example at 2.0 m spacing between rows and an average distance covered in a minute of, say 60 metres:

Para 6a entry point 5100

Basic time to treat 1 ha $= \frac{5100}{60} = 85$ minutes

If the row spacing had been 1.9 metres and mean distance covered per minute 55 metres the basic time per metre would have been $\frac{5363}{55} = 97\frac{1}{2}$ minutes.

d. Multiply the answer, calculated as in para c above, by the appropriate 'other work' and 'rest' factor (para 6b) to get overall time for application per hectare, e.g.

   i. Band application of herbicide by knapsack sprayer at 2.0 m spacing between rows and walking speed of 60 m per minute.

   ii. Basic time (see para c above) = 85 minutes.

   iii. Other work and rest factors (see para 6.b.) = 1.34.

   iv. Overall time = 85 × 1.34 = 113.9 minutes per hectare.

e. Check whether or not paragraph 7 allowances are appropriate. If so add calculated allowances or multiply overall time by specified percentage addition or additions to get target time per hectare. Examples of calculations are given in Appendix I.
IMPORTANT NOTE ALSO SEE APPENDIX II.

6. *a. Basic minutes per hectare*

Distance between rows (m)	Basic minutes per hectare at walking speed of one metre per minute
1.0	10100
1.1	9191
1.2	8433
1.3	7792
1.4	7242
1.5	6767
1.6	6350
1.7	5982
1.8	5656
1.9	5363
2.0	5100
2.1	4862
2.2	4645
2.3	4448
2.4	4267
2.5	4100

2.6	3946
2.7	3804
2.8	3671
2.9	3548
3.0	3433
4.0	2600
5.0	2100
6.0	1767

b. Additional factor (for other work, personal needs and rest).

i. Knapsack 1.34

ii. ULVA 1.31

iii. Microfit Herbi/Herbi 1.28

7. *Modification of overall times*

a. Additional items of protective clothing and equipment.
Add the following percentage for additional protective clothing and equipment where there is a label requirement or described as essential in Section 10.

Item	**Knapsack Microfit Herbi/ Herbi**
Hood	0.5%
Filtering facepiece respirator	2.0%
Face shield	0.5%

Notes:
These allowances are built into para 6.b. for the ULVA.
Allowance for normal protective clothing and equipment is included in Para 6b.

b. Number of fills required per hectare:

i. Overall application: Knapsack, Microfit Herbi/Herbi or ULVA

$$\text{Number of fills per hectare} = \frac{\text{Volume rate (l/ha)}}{\text{Applicator capacity (l)}}$$

ii. Band application: Knapsack or Microfit Herbi/Herbi

Number of fills per hectare =

$$\frac{\text{Volume rate (l/ha)} \times \text{Swathe width (m)}}{\text{Distance between rows (m)} \times \text{Applicator capacity (l)}}$$

iii. Mixing pesticide, filling, changing and refilling bottles.

c. Add the following minutes per mix, fill, change, etc.

Method	Knapsack	Microfit Herbi/ Herbi	ULVA
Mix herbicide and fill	5.81	–	–
Fill bottle with pre-mix and change	–	3.10	2.39
Change pre-filled bottle	–	2.67	1.76

APPENDIX I

Calculation of target times

Example 1

Overall application with knapsack.

Walking speed 46 m/min
Nozzle swathe width 1.4 m
Volume rate 125 l/ha
Mix herbicide and fill knapsack
No additional protective clothing required.

Overall time per hectare

$$= \frac{7242 \text{ BMs (Para 6a 1.4 m row distance)}}{46 \text{ m/min}} \times 1.34 \text{ (Para 6b)}$$

= 210.96 minutes per ha.

Add number of fills per hectare (Para 7b i) $= \frac{125 \text{ l}}{15 \text{ l}} = 8.33$

8.33 fills at 5.81 minutes per fill (Para 7c)	48.40 minutes
*Target time per hectare	259.36 minutes

Example 2

Band application with Microfit Herbi

Walking speed 60 m/min
Nozzle swathe width 1.9 m
Volume rate 20 l/ha
Swathe width 1.2 m
Fill bottle with pre-mix and change
Additional protective equipment, face shield.

Overall time per hectare

$$= \frac{5363 \text{ BMs (Para 6a 1.9 m row distance)}}{60 \text{ m/min}} \times 1.28 \text{ (Para 6b)}$$

= 114.41 minutes per ha.

Add 0.5% for face shield (Para 7a) 0.57 minutes

Add number of fills per hectare (Para 7b ii) $= \frac{20 \text{ l/ha} \times 1.2 \text{ m}}{1.9 \text{ m} \times 2.5 \text{ l}} = 5.05$

5.05 fills at 3.10 minutes per fill (Para 7c)	15.66 minutes
*Target time per hectare	130.64 minutes

* *Note:* Can be taken to be equivalent to standard time in this instance.

APPENDIX II

IMPORTANT NOTE

1. This guide covers band or overall application by KNAPSACK SPRAYERS, ULVA, MICROFIT HERBI/HERBI.

2. To use the guide it is necessary to measure the average distance covered in one minute by someone wearing requisite protective clothing, equipped with applicator full with herbicide (or water), applying 'herbicide'. Basic time so determined is then increased by the appropriate 'other work' and 'rest' allowance to give overall time per hectare. Application of para 7 amendments then give TARGET time per ha.

3. It is important to remember that time calculated as outlined above is specific to the job being done at the speed sustained during this observed minute. *If the correct level of calibrated application is to be obtained work must proceed at that speed.* Moving too quickly will give underdosage whilst slow movement will give overdosage. In the former case there will be surplus herbicide when the area has been treated, and in the latter herbicide supplied will be exhausted before the area is completely treated. Salient factors of quality control are therefore:

a. Is the treated area appropriate for the time spent on the job? In gross comparison 'target' time multiplied by area treated should equal elapsed productive time. Note however that in such gross comparison the area treated and time taken should be large enough for 'other work' to be properly represented and the appropriate amount of rest to be taken. In the comparison over short time spans, where someone may be timed over a row and the distance checked, it is important that the timed operation excludes 'rest' and 'other work' and that the comparison is in basic time, i.e. target time *divided* by the appropriate 'other work and rest' allowance. (Para 7 allowances should be deducted before dividing by the other work and rest allowances).

b. Has the specified quantity of herbicide been used?

11.12 Output guide – spot application (Weeding 2, Jan 1990)

Herbicide or insecticide spot application

Application by forestry spot gun (insecticides/herbicides) and herbicides applied by weedwiper, or granules applied by 'Pepperpot'.

1. *Conditions*

This guide applied to application of insecticides or herbicides under the following conditions:

a. Pre- and post-planting.

b. All terrain and ground conditions.

c. An adequate supply of pre-bottled or pre-mixed insecticide or herbicide.

The above should be distributed on site to avoid excessive walking to refill applicator (with 30 metres of spraying site).

d. Vegetation and crop height suitable for the application method.

e. Times are per plantation hectare.

f. Row spacing between 1.0 m and 3.0 m.

g. Trees either visible or not visible.

2. *Job specification*

The times are for the following work:

a. Applying the prescribed volume rate through the selected applicator.

b. General preparation, on/off protective clothing and equipment, wash down clothing, walk to and from rest camp, check calibration, maintain and wash out applicator.

c. Walk to and from the supply point to refill applicator (para 7).

3. *Tools and equipment*

a. Forestry spot gun, weedwiper, or Pepperpot.

b. Tools for maintenance and calibration as listed against the corresponding applicator in Section 11 of this book.

c. Equipment required for filling bottles with pre-mixed pesticide, a funnel.

d. First Aid Kit as per FASTCo Safety Guide 34 (see Section 3.2) – each operator to have a personal Emergency Kit A in the immediate work area and for there also to be a squad Emergency Kit B available on work site.

e. Protective clothing and equipment in accordance with Section 10 of this book.

f. Washing facilities for personal hygiene and wash down of protective clothing and equipment.

4. *Allowances*

The following allowances are included in the additional factor.

Applicator	Other work allowance	Rest allowance needs allowance	Overall allowance factor
Spot gun	9%	15%	1.25
Weedwiper	9%	13%	1.23
Pepperpot	8%	15%	1.24

5. *Method of using output guide*

a. On site, measure distance between rows.

b. On site, measure distance between trees in row.

c. On site, wearing the required protective clothing and equipment and with the applicator full of pesticide mixture, water or granules, walk and apply the pesticide in the prescribed manner for one minute whilst working at 'piecework' rate. Measure distance walked. If ground conditions or slope vary over the site, it will be necessary to repeat this operation a number of times to achieve an average distance covered per minute.

d. To calculate the 'Standard' time for application, divide the 'BMs' (para 6.a. against the appropriate row spacing), by the average distance covered per minute on site and multiply the additional factor (para 6.b.).

e. To calculate 'SMs' for filling bottle or changing pre-filled bottles refer to para 7.3. For the equation to give the number of fills required per hectare see paras 7.b. c. and d. The result is multiplied by the 'SMs' per fill para 7.e. to give time necessary for refilling, etc.

Examples are given in Appendix I.

6. *Standard times*

a. Basic minutes per hectare.

Distance between rows (m)	Basic minutes (BM) per hectare at walking speed of one metre per minute
1.0	10100
1.1	9191
1.2	8433
1.3	7792
1.4	7242
1.5	6767
1.6	6350
1.7	5982
1.8	5656
1.9	5363
2.0	5100
2.1	4862
2.2	4645
2.3	4448
2.4	4267
2.5	4100
2.6	3946
2.7	3804
2.8	3671
2.9	3548
3.0	3433

b. Additional factor (for other work, personal needs and rest).

Forestry spot gun	1.25
Weedwiper	1.23
Pepperpot	1.24

7. *Modification to standard times*

a. Additional items of protective clothing and equipment.
Add the following percentage for additional protective clothing and equipment where there is a label requirement or described as essential in Section 10.

Item	Knapsack Microfit Herbi/ Herbi
Hood	0.5%
Filtering facepiece respirator	2.05%
Face shield	0.5%
Cascade/Gore-Tex jacket	1.0%

Note:
Other items of protective clothing and equipment are included in the allowances listed in paras 4 and 6b.

b. Spot gun number of fills required per hectare at 5 ml dose =

$$\frac{10}{\text{Distance between rows (m)} \times \text{Plant spacing (m)}}$$

c. Weedwiper

Number of fills per hectare =

$$\frac{\text{Area of spot } (m^2) \times \text{Volume rate (l/ha)}}{\text{Distance between rows (m)} \times \text{Plant spacing} \times 0.3251}$$ See note 1

d. Pepperpot

Number of fills per hectare =

$$\frac{\text{Area of spot } (m^2) \times \text{Volume rate (kg/ha)}}{\text{Distance between rows (m)} \times \text{Plant spacing} \times 3.0}$$ See note 2

Notes:

1. 0.325 litre is the capacity of the weedwiper from the minimum mark.
2. 3.0 kg is approximate weight of contents of thirteen pepperpots.

e. Mixing pesticide, filling, change and refilling bottles.

Add the following Standard Minutes per mix, fill, change, etc.

Method	Spot gun SMs	Weedwiper SMs	Pepperpot SMs
Fill with pre-mixed	–	5.23	–
Fill bottle with pre-mix and change	6.04	–	–
Change pre-filled bottle	5.62	–	–
Refill belt with 13 pots	–	–	7.26

APPENDIX III

Example 1 Spot gun

Overall application 5 ml dose

Walking speed 48 metres/minute
Distance between rows 2.0 m
Distance between trees in row 1.9 m
Additional protective clothing, Cascade/Gore-Tex jacket
Pre-filled bottles

SMs per hectare

$= \dfrac{5100 \text{ BMs}}{48 \text{ m/min}}$ (Para 6a 2.0 m row distance) × 1.25 (Para 6b)

= 132.81 SMs

Add 1% for Cascade/Gore-Tex jacket (Para 7a)	1.33 SMs
Add number of fills per hectare at 5 ml dose (Para 7b)	
$\dfrac{10}{20 \text{ m} \times 1.9 \text{ m}} = 2.63$	
2.63 fills at 5.62 minutes per fill (Para 7e)	14.78 SMs
Total SMs per hectare	148.92 SMs

Example 2 Weedwiper

Spot treatment of 1 m diameter (0.79 m^2)

Application rate 12.5 litres per treated hectare
Walking speed 35 metres/minutes
Distance between rows 2.0 m
Distance between trees in row 2.0 m
Additional protective clothing, Cascade/Gore-Tex jacket 1%

Basic minutes per ha $= \dfrac{5100}{35}$ (Para 6a 2.0 m row spacing)

= 145.7 BM/ha.

OW and rest factor (para 6.b.) 1.23

Standard Time = 145.7 × 1.23	= 179.2 SM/ha
Add 1% for Cascade/Gore-Tex jacket	1.8 SM/ha
Weedwiper fills/ha (para 7.c.)	
$\dfrac{0.79 \times 1.25}{2.0 \times 2.0 \times 0.3251} = 7.6$	
7.6 fills at 5.23 minutes per fill (Para 7e)	39.8 SM/ha
Total	220.8 SM/ha

Example 3

Spot application with Pepperpot

Walking speed 30 m/min
Distance between rows 2.1 m
Plant spacing 2.2. m
Volume rate 38 kg/ha
Spot diameter 1.1 m (area = 0.95 m^2)
Replace full pepperpots in belt
Additional protective clothing, Cascade/Gore-Tex jacket

SMs per hectare = $\frac{4862 \text{ BMs}}{30 \text{ m/min}}$ (Para 6a 2.1 m row distance)

× . . . 1.24 (para 6b)	200.96 SMs
Add 1% for Cascade/Gore-Tex jacket (para 7a)	2.00 SMs

Add number of fills of belt per hectare (paras 7d and e) =

$$\frac{0.95 \text{ m}^2 \times 38.0 \text{ kg/ha}}{2.1 \text{ m} \times 2.2 \text{ m} \times 3.0 \text{ kg}} = 2.60$$

2.6 fills at 7.26 SMs (Para 7e)	18.88 SMs
Total SMs per hectare	221.84 SMs

12 Lists of herbicides and manufacturers

12.1 Products with full forestry label approval

Active ingredient	*Product (manufacturer)*
ammonium sulphamate	Amcide (BH&B) Root-out (Dax Products)
asulam	Asulox (RP Environmental)
atrazine	Atlas atrazine (Atlas) Unicrop Flowable Atrazine (Unicrop)
2,4-D	Dicotox Extra (RP Environmental) MSS 2,4-D Ester (Mirfield)
2,4-D/dicamba/triclopyr	Broadshot (Cyanamid)
dalapon/dichlobenil	Fydulan G (Nomix-Chipman) *Note:* stocks in existence at time of writing, but manufacture has ceased.
dicamba	Tracker (PBI)
diquat/paraquat	Farmon PDQ (Farm Protection) Parable (Zeneca)
fosamine ammonium	Krenite (Du Pont) *Note:* stocks in existence at time of writing, but manufacture has ceased.
glufosinate-ammonium	Challenge (Hoechst) Harvest (Hoechst)
glyphosate	Barclay Gallup (Barclay) Barclay Gallup Amenity (Barclay) Clayton Glyphosate (Clayton) Clayton Swath (Clayton) Glyfos (Cheminova) Glyphogan (PBI) Glyphosate 360 (Top Farm) Hilite (Nomix-Chipman) – CDA formulation Helosate (Helm) Outlaw (Barclay) Portman Glyphosate 360 (Portman) Roundup (Monsanto) Roundup (Schering/Agro Evo) Roundup Biactive (Monsanto)

	Roundup Biactive Dry (Monsanto)
	Roundup Pro Biactive (Monsanto)
	Stacato (Unicrop)
	Stefes Glyphosate (Stefes)
	Stefes Kickdown 2 (Stefes)
	Stetson (Monsanto)
	Stirrup (Nomix-Chipman) – CDA formulation
imazapyr	Arsenal 50F (Nomix-Chipman)
isoxaben	Gallery 125 (DowElanco)
	Flexidor 125
paraquat	Gramoxone 100 (Zenecca/Schering/Agro Evo)
	Scythe LC (Cyanamid)
propyzamide	Headland Judo (Headland)
	Kerb Flo (PBI Rohm+Haas)
	Kerb 50W(PBI, Rohm+Haas)
	Kerb Granules (PBI, Rohm+Haas)
triclopyr	Garlon 4 (DowElanco)
	Timbrel (DowElanco)
	Chipman Garlon 4 (Nomix-Chipman)

12.2 Products with full farm forestry label approval

Active ingredient	*Product (manufacturer)*
propaquizifop	Falcon 100 (Cyanamid)
	Shogun 100 EC (Ciba-Geigy)

12.3 Products with forestry off-label approval

Active ingredient	*Product (manufacturer)*
clopyralid	Dow Shield (DowElanco)

12.4 Products with farm forestry off-label approval

Active ingredient	*Product (manufacturer)*
cyanazine	Fortrol (Cyanamid)
fluazifop-	Fusilade 5 (Zeneca)
p-butyl	Fusilade 250 EW (Zeneca)
metazachlor	Butisan S (BASF)
pendimethalin	Stomp (Cyanamid)

12.5 Herbicides used in forestry which have approvals expired or have been commercially withdrawn during the last two years

Active ingredient	*Product (manufacturer)*
atrazine	Ashlade 4% At Gran (Ashlade) Ashlade atrazine (Ashlade) Atraflow (RP Environ) MSS atrazine 50FL (Mirfield) MSS atrazine 4G (Mirfield) MSS atrazine 80WP (Mirfield) Gesaprim 500FW (Ciba-Geigy)
atrazine/ cyanazine	Holtox (Shell)
atrazine/ dalapon	Atlas Lignum Granules (Atlas Interlates)
atrazine/ terbuthylazine	Gardoprim A 500FW (Ciba-Geigy)
benazolin/ clopyralid	Benazolox (Schering)
clopyralid/ cyanazine	Coupler SC (Shell)
clopyralid/ triclopyr	Grazon 90 (DowElanco)
2,4-D	Silvapron D (BP) BH 2,4-D Ester 50 (Rhone Poulenc)
dalapon/ di-l-menthane	Volunteered (Mandops)
dicamba	Tracker (Shell)
glyphosate	Roundup (Monsanto) Roundup Pro
glyphosate/ simazine	Rival (Monsanto)
hexazinone	Velpar (Selectokil)
isoxaben	Flexidor (DowElanco) Tripart Ratio
paraquat	Dextrone X (Nomix-Chipman)

Note:
Users normally have 2 years to utilise existing stocks of products that have been commercially withdrawn, or whose approvals have been revoked. Limited stocks of some of the products on this list may still be available – check with the supplier that they still have approval to sell these products, and that an approval still exists to apply the herbicide. If any doubt arises contact the Pesticides Safety Directorate, Mallard House, Kings Pool, 3 Peasholme Green, York, YO1 2PX. Telephone: 01904 640500.

12.6 Products with approval for use in or near water for aquatic weed control

Active ingredient	*Product (manufacturer)*
asulam	Asulox (Rhone Poulenc)
2,4-D	Atlas 2,4-D (Atlas)
	Dormone (Rhone Poulenc)
dalapon/dichlobenil	Fydulan G (Nomix-Chipman)
dichlobenil	Casoron G (Zeneca/Nomix-Chipman)
	Casoron G-SR
diquat	Midstream (Zeneca)
	Reglone (Zeneca)
fosamine ammonium	Krenite (Du Pont)
glyphosate	Barclay Gallup Amenity (Barclay)
	Glyphogan (PBI)
	Helosate (Helm)
	Roundup (Monsanto)
	Roundup Biactive Dry (Monsanto)
	Roundup Pro (Monsanto)
	Roundup Pro Biactive (Monsanto)
	Spasor (Rhone Poulenc)
	Stirrup (Nomix-Chipman)
maleic hydrazide	Bos MH 180 (Uniroyal)
	Regulox K (Rhone Poulenc)
terbutryn	Claroson 1FG (Ciba-Geigy)

Note:
Not all these products have forestry approval.
Refer to Section 3.6

12.7 Additives mentioned in the text

Product	*Manufacturer*
Agral	Zeneca
Mixture B	Service
Red dye	Hortichem
Methyl violet Garr	BDH

Note:
This is not a comprehensive list of all additives that may be used – see Section 2.6

12.8 Product suppliers

Atlas: ***Atlas Interlates Ltd.***

PO Box 38, Low Moor,
Bradford, W. Yorks. BD12 0JZ
Tel: 01274 671267
Fax: 01274 691482

Barclay: ***Barclay Chemicals (UK)***

28 Howard Street
Glossop
Derbyshire SK13 9DD
Tel: 01457 853386
Fax: 01457 853557

BASF: ***BASF plc***

Agricultural Division
Lady Lane, Hadleigh
Ipswich
Suffolk IP7 6BQ
Tel: 01473 822531
Fax: 01473 827450

BDH: ***BDH Chemicals Ltd.***

Freshwater Road
Dagenham
Essex RM18 1RZ
Tel: 01202 745520

B H & B: ***Battle, Hayward & Bower Ltd.***

Victoria Chemical Works
Croft Drive, Allenby Road Industrial Estate
Lincoln LN3 4NP
Tel: 01522 529206/541241
Fax: 01522 538960

BP: ***BP Chemicals Ltd.***

76 Buckingham Palace Road
London SW1W 0SU
Tel: 0171-581 6171
Fax: 0171-581 6387

Cheminova: ***Cheminova Agro UK Ltd.***

27 Marlow Road
Maidenhead
Berkshire SL6 7AE
Tel: 01628 770030
Fax: 01628 784215

Ciba-Geigy: ***Ciba-Geigy Agriculture***

Whittlesford
Cambridge CB2 4QT
Tel: 01223 833621
Fax: 01223 835211

Clayton: ***Clayton Plant Protection (UK) Ltd.***

97 Park Avenue
Castle Knock
Dublin 15
Republic of Ireland
Tel: 00 3531 8210127
Fax: 00 3531 8217747

Cyanamid: ***Cyanamid of Great Britain Ltd.***

Crop Protection Division
Cyanamid House
Fareham Road, PO Box 7
Gosport
Hants. PO13 0AS
Tel: 01329 224000
Fax: 01329 224335

Dax: ***Dax Products Ltd.***

76 Cyprus Road
Nottingham
NG6 5ED
Tel: 0115 960 9996

DowElanco: ***DowElanco Ltd.***

Latchmore Court
Brand Street
Hitchen
Herts. SG5 1HZ
Tel: 01462 457272
Fax: 01462 453906

Du Pont: ***Du Pont (UK) Ltd.***

Agricultural Products Department
Wedgewood Way
Stevenage
Herts. SG1 4QN
Tel: 01438 734000
Fax: 01438 734154

Farm Protection: ***Farm Protection***

Zeneca Crop Protection, UK Sales
Fernhurst, Haslemere
Surrey GU27 3JE
Tel: 01428 656564
Fax: 01428 657385

Headland: ***Headland Agrochemicals Ltd.***

Norfolk House
Great Chesterford Court
Great Chesterford
Essex CB10 1PF
Tel: 01799 530146
Fax: 01799 530229

Helm: ***Helm Great Britain Chemicals Ltd.***

Wimbledon Bridge House
1 Hartfield Road
London SW19 3RU

Hoechst: ***Hoechst UK Ltd./Agro Evo UK Crop Protection Ltd.***

Agriculture Division
East Winch Hall
East Winch, King's Lynn
Norfolk PE32 1HN
Tel: 01553 841581
Fax: 01553 841090

Hortichem: ***Hortichem Ltd.***

1 Edison Road
Churchfields Industrial Estate
Salisbury, Wilts. SP2 7NU
Tel: 01722 320133
Fax: 01722 326799

Makhteshim Agan: ***Makhteshim Agan UK Ltd.***

Forum House
45–51 Brighton Road
Redhill
Surrey RH1 6YS

Mandops: ***Mandops (UK) Ltd.***

36 Leigh Road
Eastleigh
Hants. SO5 4DT
Tel: 01703 641826
Fax: 01703 629106

Mirfield: ***Mirfield Sales Services Ltd.***

Moorend House
Moorend Lane, Dewsbury
W. Yorks. WF13 4QQ
Tel: 01924 409782
Fax: 01924 410792

Monsanto: ***Monsanto plc***

Thames Tower, Burleys Way
Leicester LE1 3TP
Tel: 0116 262 0864
Fax: 0116 253 0320

Nomix-Chipman: ***Nomix-Chipman Ltd.***

Portland Building
Portland Street
Staple Hill
Bristol BS16 4PS
Tel: 0117 957 4574
Fax: 0117 956 3461

PBI: ***Pan Britannica Industries Ltd.***

Britannica House
Waltham Cross
Herts. EN8 7DY
Tel: 01992 23691
Fax: 01992 26452

Portman: ***Portman Agrochemicals Ltd.***

Apex House, Grand Arcade
Tally-Ho Corner
North Finchley
London N12 0EH
Tel: 0181-446 8383
Fax: 0181-445 6045

Rohm & Haas: ***Rohm & Haas (UK) Ltd.***

Lennig House
2 Masons Avenue
Croydon
Surrey CR9 3NB
Tel: 0181-667 9964
Fax: 0181-686 9447

RP Environmental: ***Rhone Poulenc Environmental Products***

Regent House, Hubert Road
Brentwood
Essex EM1 4TZ
Tel: 01277 301301
Fax: 01277 260621

Sandoz: ***Sandoz Crop Protection Ltd.***

Norwich Union House
16/18 Princes Street
Ipswich
Suffolk IP1 1QT
Tel: 01473 255972
Fax: 01473 258252

Schering: ***Schering Agriculture***

Nottingham Road
Stapleford
Nottingham NG9 8AJ
Tel: 0115 939 0202
Fax: 0115 939 8031

Service: ***Service Chemicals Ltd.***

Lachester Way
Royal Oak Industrial Estate
Daventry, Northants. NN11 5PH
Tel: 01327 704444
Fax: 01327 71154

Shell: ***Shell Chemicals UK Ltd.***

Agricultural Division
Heronshaw House
Ermine Business Park
Huntingdon
Cambs. PE18 6YA
Tel: 01480 414140
Fax: 01480 444444

Stefes: ***Stefes Plant Protection Ltd.***

Huntingdon Business Centre
Blackstone Road
Huntingdon
Cambs. PE19 6EF
Tel: 01480 435101
Fax: 01480 420041

Top Farm: ***Top Farm Formulations Ltd.***

115 Carrowreagh Road
Garvagh, Coleraine
Co. Londonderry BT51 5LQ
N. Ireland
Tel: 01266 557075
Fax: 01266 557012

Tripart: ***Tripart Farm Chemicals Ltd.***

Swan House, Beulah Street
Gaywood, King's Lynn
Norfolk PE30 4ND
Tel: 01553 674303
Fax: 01553 674422

Unicrop: ***Universal Crop Protection Ltd.***

Park House, Cookham
Maidenhead
Berks. SL6 9DS
Tel: 01628 526083
Fax: 01628 810457

Uniroyal: ***Uniroyal Chemical Ltd.***

Kennet House
4 Langley Quay
Slough
Berks. SL3 6EH
Tel: 01753 580888

Zeneca: ***Zeneca Crop Protection (formerly ICI Agrochemicals Garden & Professional Products)***

Fernhurst
Haslemere
Surrey GU27 3JE
Tel: 01428 656564
Fax: 01428 657385

12

13 Sources of advice

13.1 Forestry Authority Offices

Policy and general operational advice

The Forestry Authority Scotland

Scotland National Office

Portcullis House
21 India Street
Glasgow G2 4PL
Tel: 0141 248 3931

Conservancies Scotland

Dumfries and Galloway Conservancy

134 High Street
Lockerbie
Dumfriesshire
DG11 2BX
Tel: 01576 202858

Grampian Conservancy

Ordiquhill
Portsoy Road
Huntly AB54 4SJ
Tel: 01466 794542

Highland Conservancy

Hill Street
Dingwall
Ross-shire IV15 9JP
Tel: 01349 62144

Lothian and Borders Conservancy

North Wheatlands Mill
Wheatlands Road
Galashiels TD1 2HQ
Tel: 01896 550222

Perth Conservancy

10 York Place
Perth PH2 8EP
Tel: 01738 442830

Strathclyde Conservancy

21 India Street
Glasgow G2 4PL
Tel: 0141 248 3931/041 221 2506

The Forestry Authority England

England National Office

Great Eastern House
Tenison Road
Cambridge CB1 2DU
Tel: 01223 314546

Conservancies England

Cumbria and Lancashire Conservancy

Peil Wyke
Bassenthwaite Lake
Cockermouth
Cumbria CA13 9YG
Tel: 01768 776616

East Anglia Conservancy

Santon Downham
Brandon
Suffolk IP27 0TJ
Tel: 01842 815544

East Midlands Conservancy

Willingham Road
Market Rasen
Lincs. LN8 3RQ
Tel: 01673 843461/842644

Greater Yorkshire Conservancy

Wheldrake Lane
Crockey Hill
York YO1 4SG
Tel: 01904 448778

Hampshire and West Downs Conservancy

Alice Holt
Wrecclesham
Farnham
Surrey GU10 4LF
Tel: 01420 23337
Fax: 01420 22988

Kent and East Sussex Conservancy

Furance Lane
Lamberhurst
Tunbridge Wells
Kent TN3 8LE
Tel: 01892 891100

Northumberland and Durham Conservancy

Redford
Hamsterley
Bishop Auckland
Co Durham DL13 3NL
Tel: 01388 88721

Thames and Chilterns Conservancy

The Old Barn
Upper Wingbury Farm
Wingrave
Aylesbury
Bucks. HP22 4RF
Tel: 01296 681181/681381

West Country Conservancy

The Castle
Mamhead
Exeter
Devon EX6 8HD
Tel: 01626 890666

West Midlands Conservancy

Rydal House
Colton Road
Rugeley
Staffs. WS15 3HF
Tel: 01889 585222

Wye and Avon Conservancy

Bank House
Bank Street
Coleford
Glos. GL16 8BA
Tel: 01594 810983

The Forestry Authority Wales

Wales National Office

North Road
Aberystwyth
Dyfed
SY23 2EF
Tel: 01970 625866

Conservancies Wales

Mid Wales Conservancy

The Gwalia
Llandrindod Wells
Powys LD1 6AA
Tel: 01597 825666

Gwarchodfa Canolbarth Cymru

Y Gwalia
Llandrindod Wells
Powys LD1 6AA
Ffon: 01597 825666

North Wales Conservancy

Clawdd Newydd
Ruthin
Clwyd
LL15 2NL
Tel: 01824 750492/493

Gwarchodfa Gogledd Cymru

Clawdd Newydd
Rhuthun
Clwyd
L15 2NL
Ffon: 01824 750492/493

South Wales Conservancy
Cantref Court
Brecon Road
Abergavenny
Gwent NP7 7AX
Tel: 01873 850060

Gwarchodfa de Cymru

Cwrt Cantref
Heol Aberhonddu
Y Fenni
Gwent NP7 7AX
Ffon: 01873 850060

The Forestry Commission Research Division

Silviculture (South)
Herbicides in lowland forests
Alice Holt Lodge, Wrecclesham, Farnham, Surrey
GU10 4LH
Tel: 01420 22255

Silviculture (North)
Herbicides in upland forests
Northern Research Station, Roslin, Midlothian EH25 9SY
Tel: 0131-445 2176

The Forestry Commission Technical Development Branch

Protective clothing and appliances
Branch Headquarters
Ae Village
Dumfries
DG1 1QB
Tel: 01387 860264

The Forestry Commission
Education, Safety and Training Branch

FC Education, Safety and Training Policy
231 Corstorphine Road
Edinburgh EH12 7AT
Tel: 0131 334 0303

13.2 Other sources of advice

Pesticides, health and safety

The Health and Safety Executive (HSE)
HSE Information Centre
Broad Lane
Sheffield S3 7HQ
Tel: 0114 289 2345
Fax: 0114 289 2333

All aspects of pesticide legislation

The Pesticides Safety Directorate
Mallard House, Kings Pool
3 Peasholme Green
York Y01 2PX
Tel: 01904 640500
Fax: 01904 455733

13

14 Glossary and abbreviations

14.1 Abbreviations used in the text

14.1.1 Species

	Common name	Botanical name
CP	Corsican pine	*Pinus nigra* var. *maritima*
DF	Douglas fir	*Psuedotsuga menziesii*
EL	European larch	*Larix decidua*
GF	Grand fir	*Abies grandis*
JL	Japanese larch	*Larix kaempferi*
LC	Lawson cypress	*Chamaecyparis lawsoniana*
LP	Lodgepole pine	*Pinus contorta*
NF	Noble fir	*Abies procera*
NS	Norway spruce	*Picea abies*
OMS	Serbian spruce	*Picea omorika*
SP	Scots pine	*Pinus sylvestris*
SS	Sitka spruce	*Picea sitchensis*
WH	Western hemlock	*Tsuga heterophylla*
WRC	Western red cedar	*Thuja plicata*

14.1.2 Other abbreviations and symbols

a.e.	acid equivalent
a.i.	active ingredient
bm	basic minutes
cm	centimetre(s)
FA	Forestry Authority (part of the Forestry Commission)
FASTCo	Forestry and Arboriculture Safety and Training Council
FC	Forestry Commission
FE	Forest Enterprise (part of the Forestry Commission)
g	gramme(s)
ha	hectare
HSE	Health and Safety Executive
HV	high volume
kg	kilogramme(s)
kPa	kilopascals
l	litre(s)
LV	low volume
m	metre(s)

MAFF	Ministry of Agriculture, Fisheries and Food
ml	millilitre(s)
mm	millimetre(s)
mph	miles per hour
MV	medium volume
NRS	Northern Research Station (FC Research Division)
PSD	Pesticides Safety Directorate
p.s.i.	pounds per square inch
rpm	revolutions per minute
s or sec	second(s)
s.c.	suspension concentrate
sm	standard minute
ULV	ultra low volume
VLV	very low volume
VMD	volume median diameter
w.p.	wettable powder
w/v	weight per volume (weight active ingredient/unit volume product)
w/w	weight per weight (weight active ingredient/unit weight product)

14.2 Glossary of general and technical terms

Acid equivalent (a.e.)

The amount of active ingredient expressed in terms of parent acid.

Active ingredient (a.i.)

That part of a pesticide formulation from which the phytotoxicity (weedkilling effect) is obtained.

Additive

A non-herbicidal material which is added to a herbicide formulation to improve its performance in any way.

Adjuvant

A non-herbicidal material which is added to a herbicide formulation to enhance the phytotoxicity (killing effect) of the formulation.

Agitation

Continual mixing of a liquid preparation of a pesticide (usually at the state of final dilution) by shaking or stirring.

Application method (or pattern)

The arrangement of areas which receive an application of herbicide and the relationship of this pattern to the crop trees (when present). There is a close association between such a pattern and the application equipment used: the term 'application method' should strictly speaking refer to the applicator and its use as well as the application pattern produced. Sub-divisions are:

Band application

Herbicide applied to a strip of ground or vegetation, normally centred on a row of crop trees.

Directed application

The herbicide spray is directed to hit target weeds and to avoid the crop trees.

Guarded application

A spray where the crop trees are physically protected from direct contact with the herbicide by a guard or guards, usually attached to the applicator.

Incremental drift application

A form of application where the herbicide is sprayed as droplets small enough to be wind-assisted to their target and applied in successive overlapping bands so that a relatively even coverage of the whole area is achieved.

Spot application

Herbicide applied as individual spots to bare ground or vegetation, normally immediately around the crop area.

Stem or cut stump treatments

Herbicide applied to individual stems or cut stumps wherever they may occur on a weeding site (not necessarily over the whole site).

Overall application

Herbicide is applied over the whole weeding site.

Applicator

A piece of equipment designed to distribute herbicide on to ground or vegetation.

Approved product

A pesticide which has been approved for use by the Pesticides Safety Directorate on grounds of safety and efficacy under The Control of Pesticides Regulations 1986.

Band application

See Application method.

Calibration

The process of calculation, measurement and adjustment (of parameters such as nozzle type, operating pressure, walking speed) by which the correct application rate is achieved.

Carrier

A liquid or solid material within which a pesticide is dispersed (e.g. solution or suspension) to facilitate application.

Chlorosis

Loss of green colour in plant foliage.

Coarse grasses

An imprecise term used to describe grasses of a generally tall, bulky, rank, stiff and often tussocky nature which are also relatively more resistant to grass herbicides. By contrast, the so-called soft grasses are usually more susceptible to grass herbicides.

Contact herbicide

One that kills or injures plant tissue close to the point of contact or entry into the plant (contrast with translocated herbicide).

Controlled droplet application (CDA)

Systems where droplets are generated by separating from points on a rapidly rotating disc or cage. Droplets so generated are more uniform in size than those generated through a hydraulic spray nozzle.

Diluent

The liquid added to a herbicide concentrate to increase its volume to an extent suitable for the applicator to be used.

Direct applicator

A piece of equipment which transfers liquid herbicide to a weed by direct contact with no intervening passage of droplets in air.

Directed application

See Application method.

Dormant period

The period of the year when the aerial part of a plant is not in active growth.

Emulsifiable concentrate

See Formulation.

Emulsion

A mixture in which fine globules of one liquid are dispersed in another, e.g. oil in water.

Esters and salts

Different groups of compounds derived from an organic acid. Esters are normally oil-soluble.

Flushing

The commencement of growth of a plant above ground, characterised by sap flow and swelling and bursting of buds. Flushing follows the end of dormancy and marks the beginning of the growing season.

Formulation

a. The process of preparing a pesticide in a form suitable for practical use either neat or after dilution.
b. The material resulting from the above process.

Types of formulation are:

Emulsifiable concentrate

A concentrated solution of a herbicide and an emulsifier in an organic solvent which will form an emulsion on mixing with water.

Granules

A free flowing dry preparation of herbicide (in a solid carrier in the form of particles within a given diameter range) which is ready for use.

Liquid

A concentrated solution of herbicide which mixes readily with water.

Suspension concentrate

A stable suspension of a solid herbicide in a fluid, intended for dilution before use.

ULV formulation

Herbicide in a special blend of oils intended for application through a rotary atomiser without dilution.

Wettable powder

Herbicide in a powder so formulated that it will form a suspension when mixed with water.

Herbi

Trade name of a controlled droplet applicator.

Herbicide

A chemical which can kill or damage plants.

Hormonal action

The mode of certain herbicides (e.g. 2,4-D) which achieve their effect by interfering with the growth regulator mechanisms of the weed plant. Bending, curling and deformation (epinasty) of shoots and leaves is a common symptom of such effects.

Incremental drift

See Application method.

Lammas growth

A second flush of growth that can occur in some species after predetermined growth has ceased, in July/August.

Liquid formulation

See Formulation.

Low-volatile ester

An ester of an organic acid (e.g. 2,4-D) which has a sufficiently long chain of carbon atoms in the molecule to reduce the amount that evaporates during and after spraying to an insignificant level.

Low volume

See Volume rate.

Medium volume

See Volume rate.

Moderately resistant

Control from the herbicide application may be inadequate.

Moderately susceptible

Control from the herbicide application should be adequate, but weeds may re-grow/re-invade.

Moderately tolerant

Some damage or check from a herbicide is likely, but the crop species should survive.

Nozzle type

Nozzles for liquid herbicide applicators are described by the spray patterns produced:

Fan

Spray droplets are emitted in a fan shape. Overlapping of the tapered edges will produce an even distribution.

Even fan

Spray droplets are emitted evenly over the width of the fan.

Hollow cone

Spray droplets are emitted in the shape of a hollow cone which produces a ring on the ground.

Solid cone
Spray droplets are distributed over the whole area of the base of the cone.
Anvil flood jet
Spray droplets are emitted in a wide fan by the stream of herbicide striking against a dispersing surface (the anvil) to produce a wide band pattern on the ground.
Solid stream
Herbicide is emitted as a continuous jet (rather like a hose-pipe) and not broken down into separate droplets.
Variable
A nozzle in which the distribution of spray can be adjusted from a narrow jet to a wide cone pattern.

Pepperpot

A simple handheld container with holes drilled in the lid for distribution of granular herbicides.

Persistence

The length of time a herbicide remains active in the soil.

Pesticide

A generic term covering herbicides, fungicides, insecticides, and defined legally in the Food and Environment Protection Act 1985.

Photosynthetic process

The series of chemical reactions in the plant leaf by which sugar is made from carbon dioxide, water and sunlight. Some herbicides achieve their effect by interfering with one or more of these reactions.

Poisons rules

Regulations governing the labelling, storage and sale of materials listed as poisons under the Poisons Act 1972. See Section 2.9.

Post-planting

After the crop has been planted.

Pre-planting

Before the crop is planted.

Product

A formulation of a herbicide of fixed (but usually confidential) composition and of known strength (% content of the active ingredient or acid equivalent) which is commercially marketed under a particular brand name. It is the individual product which is given approval under The Control of Pesticides Regulations 1986.

Product rate

The amount (weight or volume) of active ingredient or product applied per unit area, per plant, per incision, etc. Because of the range of possible meanings, ambiguity should be avoided by quoting the appropriate units, e.g. litres of product per treated hectare.

Residual herbicide

One which remains active in the soil for a period after it has been applied, and will affect weeds growing in treated soil.

Resistant

Unaffected or undamaged by exposure to a herbicide applied at a stated rate. Usually used to describe the reaction of weed species to a herbicide.

Restocking area

An area where one stand of trees has been felled and is being replaced by another.

Rotary atomiser

A herbicide applicator in which the herbicide liquid is broken into droplets of a more or less uniform diameter by being thrown from the edge of a spinning disc.

Selective herbicide

One which, if used appropriately, will kill or damage some plant species while leaving others unaffected.

Senescence

The annual ageing process by which each autumn the leaves or above ground parts of plants wither and die back.

Sensitive

Easily damaged by herbicide.

Setting bud

The formation of buds in readiness for winter dormancy. Usually follows closely after the season's rapid growth and the development of a waxy cuticle.

Soft (or fine) grasses

See Coarse grasses.

Soil-acting herbicide

One which is active through the soil, usually entering plants through the roots.

Soil classification

Most product labels use the system in MAFF Pamphlet 3001, *Soil Texture (89) system and pesticide use.* In this system

Stony soil	5–15% stones (>2mm).
Very stony soil	> 15% stones.
Calcareous soil	> 5% calcium carbonate.
Peaty soil	20–35% organic matter.
Organic soil/layer	> 30% organic matter.
Peat soil	> 35% organic matter.

Spot application

See Application method.

Standard output

Output expressed in terms of standard time taken to perform an operation.

Standard time

A measurement of time including direct work, indirect work and rest allowance. The total time required to perform an operation.

Stem treatment

See Application method.

Surfactant (or surface acting agent)

A substance which is added to a spray solution to reduce the surface tension of the liquid and increase the emulsifying, spreading and wetting properties, so enhancing the tenacity of the herbicide on the treated plant.

Susceptible

Readily controlled by a herbicide applied at a stated rate.

Suspension

Particles dispersed through (but not dissolved in) a liquid.

Suspension concentrate

See Formulation.

Swathe

A strip of ground or vegetation of a given width which receives herbicide from a single pass of an applicator.

Tillered

A description often applied to grasses to indicate shooting or branching at the base of the plant.

Tolerant

Unaffected or undamaged by exposure to a herbicide. Usually used to describe the reaction of crop trees to a selective herbicide.

Total herbicide

A herbicide used in such a way as to kill all vegetation.

Toxicity

The capacity of a material to produce any noxious effect – reversible or irreversible – on the subject referred to.

Translocated herbicide

One which is moved within the plant and can affect parts of the plant remote from the point of application.

Treated area

The area of ground or plantation that is actually covered with herbicide (usually expressed as treated hectares).

Ultra low volume

See Volume rate.

Very low volume

See Volume rate.

Volatilisation

The rapid evaporation of a liquid. With herbicides, such a process can result in a highly mobile cloud of product, which can drift and cause considerable damage to adjacent areas.

Volume median diameter

The diameter in a droplet spectrum at which half the volume of the spray is contained in smaller and half in larger droplets.

Volume rate

The amount of spray solution (diluent plus herbicide) applied per unit area. Volume rates are frequently described as high, medium, low, very low or ultra low volume but several conventions exist as to the range of rates to which each term refers. The definitions used in this publication are to be found in Section 11.2 but where precision is important it is advisable to state the exact rate.

Weed spectrum

The range of undesirable species which are killed or adequately controlled by a herbicide.

Wettable powder

See Formulation.

Wetter or wetting agent

A surfactant.

Weight/volume (w/v)

A means of expressing the amount of active ingredient in a commercial formulation by relating this amount, by weight, to the volume of the formulation. In expressing the ratio as a percentage, the assumption is made that 1.0 litre of every formulation weighs 1.0 kg (e.g. 20% w/v = 0.2 kg in every 1.0 litre of formulation).

Weight/weight (w/w)

A means of expressing the amount of active ingredient in a commercial formulation by relating this amount, weight by weight, to the weight of the formulation (e.g. 20% w/w = 0.2 kg in every 1.0 kg of formulation).

15 Index to weeds and chemicals

Abies grandis (grand fir) 77, 124
Abies procera (noble fir) 77, 83, 89, 124, 130, 171
Acer spp. (sycamore, maple) 70, 122
Achillea millefolium (yarrow) 52
Aegopodium podagraria (ground elder) 47
Agropyron repens (couch grass) 38
Agrostis spp. (bent grasses) 38
Ajuga spp. (viper's bugle) 52
alder (*Alnus* spp.) 70, 122, 153
Alnus spp. (alder) 70, 122, 153
Amcide (ammonium sulphamate) 28, 120, 139–40, 146–7, 155, 156–7, 240, 277
Ammonium sulphamate (Amcide, Root-out) 28, 120, 139–40, 146–7, 155, 156–7, 240, 277
Anagallis arvensis (scarlet pimpernel) 50
Annual meadow grass (*Poa annua*) 42
Annual mercury (*Mercurialis annua*) 43
Annual milk thistle (sow thistle) (*Sonchus asper*) 43
Annual/small nettle (*Urtica urens*) 43
Anthemis spp. (stinking chamomile) 51
Anthemis arvensis (corn chamomile) 45
Anthoxanthum odoratum (sweet vernal) 38
Anthriscus sylvestris (cow parsley) 45
Aphanes arvensis (parsley piert) 49
Arabidopsis thaliana (thale cress) 51
Arable weeds 168–207
Arrhenatherum elatius (false oat) 39, 173
Arsenal 50F (imazapyr) 28, 35, 36, 37, 87–8, 97, 107–8, 109, 118–19, 120, 134–5, 155, 163–5, 278
Artemisia vulgaris (mugwort) 49
Ash (*Fraxinus excelsior*) 70, 122
Ashlade 4% At Gran *see* atrazine
Ashlade Atrazine *see* Glyphosate
Aspen poplar (*Populus* spp.) 122
Asulam (Asulox) 28, 97, 98–100, 277, 280
Asulox (Asulam) 28, 97, 98–100, 277, 280
Atlas Atrazine (atrazine) 28, 35, 36, 37, 59–69, 277, 279
Atlas Lignum Granules (atrazine with dalapon) 279
Atlas 2,4-D *see* 2,4-D
Atraflow *see* atrazine

Atrazine (Ashlade 4% At Gran, Ashlade Atrazine, Atlas Atrazine, Atraflow, Gesaprim 500FW, MSS Atrazine 80WP, MSS Atrazine 50FL, MSS Atrazine 4G, Unicrop Flowable) 28, 35, 36, 37, 59–69, 277, 279
Atrazine with cyanazine (Holtox) 279
Atrazine with dalapon (Atlas Lignum Granules) 279
Atrazine with terbuthylazine (Gardoprim A 500FW) 279
Atriplex patula (common orache) 44

Barclay Gallup *see* Glyphosate
Barclay Gallup Amenity *see* Glyphosate
Beech (*Fagus sylvatica*) 70, 122
Bellis spp. (common daisy) 44
Benazolin with clopyralid (Benazolox) 279
Benazolox (benazolin with clopyralid) 279
Bent grasses (*Agrostis* spp.) 38
Bents 172
Betula spp. (birch) 70, 122, 153
BH 2,4-D Ester 50 *see* 2,4-D
Bilderdykia convolvulus (black bindweed) 43, 172
Birch (*Betula* spp.) 70, 122, 153
Bitter cress, hairy (*Cardamine hirsuta*) 172
Bittersweet 172
Black bindweed (*Bilderdykia convolvulus*) 43, 172
Black grass 172
Black nightshade (*Solanum nigrum*) 174
Blackthorn (*Prunus spinosa*) 122, 153
Bos MH 180 (maleic hydrazide) 280
Bracken 36, 97–110
Bramble (*Rubus* spp.) 122
Brassica napus (volunteer oilseed rape) 52
Broad dock (*Ramex obtusifolius*) 43
Broadshot (2,4-D with dicamba and triclopyr) 35, 36, 37, 79–80, 120, 121, 126–7, 148–9, 155, 158–9, 277, 280
Brome, barren 172
Broom (*Cytisus scoparius*) 122
Buckthorn (*Rhamnus cathartica*) 122
Butisan S (metazachlor) 28, 169, 170, 177, 197–201, 278
Buttercup spp. (*Ranunculus* spp.) 43
Buttercup, corn (*Ranunculus arvensis*) 44
Buttercup, creeping 172

Calamagrostis epigejos (small reed grass) 39, 85, 93
Calluna vulgaris (heather) 109–19

Canary grass, awned 172
Capsella bursa-pastoris (shepherd's purse) 50, 175
Cardamine hirsuta (hairy bitter cress) 47, 172
Cardarla draba (hoary cress) 48
Casoron G (dichlobenil) 28, 280
Castanea sativa (chestnut) 122
Cat's ear (*Hypochoeris* spp.) 43
Cedar, western red (*Thuja plicata*) 70, 77, 83, 124, 130, 171
Cerastium spp. (mouse ear) 49
Challenge (glufosinate ammonium) 28, 35, 37, 81–2, 277
Chamaecyparis lawsoniana (Lawson cypress) 83, 130
Chamerion angustifolium (willowherb) 52
Chamomile, common 172
Chamomile, stinking 172
Chamomilla recutita (scented mayweed) 50
Chamomilla suaveolens (pineapple weed) 49, 174
Charlock (*Sinapis arvensis*) 43, 172
Chenopodium album (fat hen) 46, 173
Chestnut (*Castanea sativa*) 122
Chickweed, common (*Stellaria media*) 172
Chipman Garlon 4(triclopyr) 28, 35, 36, 37, 94–6, 120, 121, 136–8, 144–5, 153–4, 155, 166–7, 278
Chrysanthemum segetum (corn marigold) 45
Cirsium arvense (creeping thistle) 46
Cirsium vulgare (spear thistle) 51
Claroson lFG (terbutryn) 280
Clayton Glyphosate *see* Glyphosate
Clayton Swath *see* Glyphosate
Cleavers (*Galium aparine*) 44, 172
Clopyralid (Dow Shield) 28, 35, 36, 70–4, 169, 170, 278
Clopyralid with benazolin (Benazolox) 279
Clopyralid with cyanazine (Coupler SC) 279
Clopyralid with triclopyr (Grazon 90) 279
Clover (from seed) 172
Clover spp. (*Trifolium* spp.) 44
Cocksfoot (*Dactylis glomerata*) 39, 93
Coltsfoot (*Tussilago farfara*) 44, 172
Common chickweed (*Stellaria media*) 44
Common daisy (*Bellis* spp.) 44
Common dandelion (*Taraxacum officinalis*) 44
Common couch 172
Common orache (*Atriplex patula*) 44
Common vetch (*Vicia sativa*) 44

Convolvulus arvensis (field bindweed) 46
Corn buttercup (*Ranunculus arvensis*) 44
Corn chamomile (*Anthemis arvensis*) 45
Corn marigold (*Chrysanthemum segetum*) 45
Corn mint (*Mentha arvensis*) 45
Corn poppy (*papaver rhoeas*) 45
Corn spurrey (*Spergula arvensis*) 45
Cornus sanguinea (dogwood) 122
Corylus avellana (hazel) 122
Couch grass (*Agropyron repens*) 38
Coupler SC (clopyralid with cyanazine) 279
Cow parsley (*Anthriscus sylvestris*) 45
Crane's-bill (*Geranium* spp.) 45, 172
Crane's-bill, cut-leaved 172
Crataegus monogyna (hawthorn) 122
Creeping bent (watergrass) 172
Creeping buttercup (*Ranunculus* repens) 45
Creeping cinquefoil (*Potentilla* repens) 45
Creeping soft grass (*Holcus mollis*) 41, 60, 93, 172
Creeping thistle (*Cirsium arvense*) 46
Crepis capillaris (smooth hawksbeard) 51
Crested dog's tail 173
Curled dock (*Rumex crispus*) 46, 173
Cyanazine (Fortrol) 28, 45, 169, 170, 177, 178–85, 278
Cyanazine with atrazine (Holtox) 279
Cyanazine with Clopyralid (Coupler SC) 279
Cypress, Lawson (*Chamaecyparis lawsoniana*) 83,130
Cytisus scoparius (broom) 122

Dactylis glomerata (cocksfoot) 39, 93
Dalapon 28
Dalapon with atrazine (Atlas Lignum Granules) 279
Dalapon with dichlobenil (Fydulan G) 35, 36, 37, 75–6, 277, 280
Dalapon with di-l-menthane (Volunteered) 279
Deadnettle, henbit (*Lamium amplexicaule*) 173
Deadnettle, red (*Lamium purpureum*) 173
Deadnettle, white 173
Deschampsia caespitosa (tufted hair grass) 39
Deschampsia flexuosa (wavy hair grass) 40, 85
Dextrone X (paraquat) 28, 278, 279
Dicamba (*Tracker*) 28, 97, 101–2, 277, 279
Dicamba with triclopyr and 2,4-D (Broadshot) 35, 36, 37, 79–80, 120, 121, 126–7, 148–9, 155, 158–9, 277, 280

Dichlobenil (Casoron G, Casoron G-SR) 28, 280
Dichlobenil with dalapon (Fydulan G) 35, 36, 37, 75–6, 277, 280
Dicotox Extra (2,4-D) *see* 2,4-D
Di-l-menthane with dalapon (Volunteered) 279
Diquat (Midstream, Reglone) 28, 280
Diquat with paraquat (Farmon, Parable) 35, 37, 277
Dock, broadleaved 173
Dock spp. (*Rumex* spp.) 46
Dogwood (*Cornus sanguinea*) 122
Dormone see 2,4-D
Dow Shield (clopyralid) 28, 35, 36, 70–4, 169, 170, 278

Elder (*Sambucus nigra*) 122
Elm (*Ulmus procera*) 122
Equisetum arvense (field horsetail) 46, 173
Euphorbia spp. (spurge) 51

Fagus sylvatica (beech) 70, 122
Falcon 100 (propaquizafop) 28, 170, 171, 206–7, 278
False oat (*Arrhenatherum elatius*) 39, 173
Farmon PDQ (*diquat with paraquat*) 35, 37, 277
Fat-hen (*Chenopodium album*) 46, 173
Festuca arundinacea (tall fescue) 40
Festuca ovina (sheep's fescue) 40
Festuca pratensis (meadow fescue) 40, 173
Festuca rubra (red fescue) 41
Field bindweed (*Convolvulus arvensis*) 46
Field horsetail (*Equisetum arvense*) 46, 173
Field pansy (*Viola arvensis*) 46
Field penny-cress (*Thlaspi arvense*) 46
Field speedwell (*Veronica persicae*) 46
Fir, Douglas (*Pseudotsuga menziesii*) 70, 75, 77, 83, 87, 89, 94, 124, 130, 134, 136, 171
Fir, grand (*Abies grandis*) 77, 124
Fir, noble (*Abies procera*) 77, 83, 89, 124, 130, 171
Fleabane, common 173
Flexidor 125 (isoxaben) 28, 35, 36, 89–90, 169, 170, 174, 278, 279
Fluazifop-p-butyl (Fusilade 5, Fusilade 250 EW) 28, 170, 186–97, 278
Fool's parsley 173
Forget-me-not, field (*Myosotis arvensis*) 47, 173
Fortrol (cyanazine) 28, 169, 170, 177, 178–85, 278
Fosamine ammonium (Krenite) 28, 120, 121, 128–9, 277, 280
Foxglove 173

Fraxinus excelsior (ash) 70, 122
Fumaria officinalis (fumitory) 47, 173
Fumitory, common (*Fumaria officinalis*) 47, 173
Fusilade 5 (Fluazifop-p-butyl) 28, 170, 186–97, 278
Fydulan G (dalapon with dichlobenil) 35, 36, 37, 75–6, 277, 280

Galeopsis tetranit (hemp nettle) 48, 173
Galinsoga parviflora (gallant soldier) 47
Galium aparine (cleavers) 44, 172
Gallant soldier (*Galinsoga parviflora*) 47
Gallery 125 (isoxaben) 28, 35, 36, 37, 89–90, 169, 170, 174, 278
Gardoprim A 500FW (atrazine with terbuthylazine) 279
Garlon 4(triclopyr) 28, 35, 36, 37, 94–6, 120, 121, 136–8, 144–5, 153–4, 155, 166–7, 278
Geranium spp. (crane's bill) 45
Gesaprim 500FW *see* atrazine
Glufosinate ammonium (Challenge, Harvest) 28, 35, 37, 81–2, 277
Glyphogan *see* Glyphosate
Glyphosate (Barclay Gallup, Barclay Gallup Amenity, Clayton Glyphosate, Clayton Swath, Glyphogan, Glyphosate-360, Helosate, Hilite, Outlaw, Portman G Glyphosate, Roundup, Roundup Biactive, Roundup Pro Biactive, Stacato, Stefes Glyphosate, Stefes Kickdown 2, Stetson, Stirrup) 3–4, 28, 35, 36, 37, 83–6, 97, 103–6, 109, 114–17, 120, 121, 130–3, 141–3, 150–2, 155, 160–2, 277–8, 279, 280
Glyphosate with simazine (Rival) 279
Golden rod (*Solidago canadensis*) 46
Gorse (*Ulex* spp.) 122
Gramoxone (paraquat) 28, 278, 279
Grasses 35–96
Grazon 90 (Clopyralid with triclopyr) 279
Greater plantain (*Plantago major*) 46
Gromwell, field 173
Ground elder (*Aegopodium podagraria*) 46
Groundsel (*Senecio vulgaris*) 46, 173
Guelder rose (*Viburnum opulus*) 123

Hairy bittercress (*Cardamine hirsuta*) 47, 172
Harvest (glufosinate ammonium) 28, 35, 37, 81–2
Hawkbit (*Leontodon autumnalis*) 47
Hawkweed spp. (*Hieracium* spp.) 48
Hawthorn (*Crataegus monogyna*) 122
Hazel (*Corylus avellana*) 122
Heather (*Calluna vulgaris*) 109–19

Helosate *see* Glyphosate
Hemlock, western (*Tsuga heterophylla*) 59, 77, 124
Hemp-nettle, common (*Galeopsis tetranit*) 48, 173
Henbit deadnettle (*Lamium amplexicaule*) 48
Heracleum sphondylium (hogweed) 48
Herbaceous broadleaved weeds 35–98
Hexazinone (Velpar) 279
Hieracium spp. (hawkweed spp.) 48
Hilite *see* Glyphosate
Hoary cress (*Cardarla draba*) 48
Hoary plantain (*Plantago media*) 48
Hogweed (*Heracleum sphondylium*) 48
Holcus lanatus (Yorkshire fog) 41, 176
Holcus mollis (Creeping Soft Grass) 41, 60, 85, 93, 172
Holtox (cyanazine with atrazine) 279
Hypercium perforatum (St John's wort) 51
Hypochoeris spp. (cat's ear) 43

Imazapyr (Arsenal 50F) 28, 35, 36, 37, 87–8, 97, 107–8, 109, 118–19, 120, 134–5, 155, 163–5, 278
Isoxaben (Gallery 125, Flexidor 125, Tripart Ratio) 28, 35, 36, 37, 89–90, 169, 170, 174, 278, 279
Ivy-leaved speedwell (*Veronica hederifolia*) 48

Japanese knotweed (*Polygonum japonicum*) 123
Juncus spp. (rush) 42

Kerb 50W (Propyzamide) 28, 35, 36, 37, 91–3, 169, 170, 177, 238, 278
Kerb Flowable (Propyzamide) 28, 35, 36, 37, 91–3, 169, 170, 177, 238, 278
Kerb Granules (Propyzamide) 28, 35, 36, 37, 91–3, 169, 170, 177, 238, 278
Knotgrass (*Polygonum aviculare*) 48, 173
Krenite (fosamine ammonium) 28, 120, 121, 128–9, 277, 280

Lamium amplexicaule (henbit deadnettle) 48
Lamium purpureum (red deadnettle) 50
Larch, European (*Larix decidua*) 59, 89
Larch, Japanese (*Larix kaempferi*) 37, 70, 89, 134, 171
Larix decidua (European larch) 59, 89
Larix kaempferi (Japanese larch) 37, 70, 89, 134, 171
Laurel (*Prunus laurocerasus*) 122
Leontodon autumnalis (hawkbit) 47
Ligustrum vulgare (privet) 122

Maleic hydrazide (Bos MH 180, Regulox K) 280
Maple (*Acer* spp.) 122
Marigold, common 174
Marigold, corn (*Chrysanthemum segetum*) 44
Mat grass 174
Matricaria perforata (scentless mayweed) 50, 174
Mayweed scented (*Chamomilla recutita*) 49, 174
Mayweed, scentless (*Matricaria perforata*) 50, 174
Meadow buttercup (*Ranunculus acris*) 48
Meadow fescue (*Festuca pratensis*) 40
Meadow foxtail 174
Meadow grass, annual 174
Meadow grass, rough 174
Meadow grass, smooth 174
Mentha arvensis (corn mint) 45
Mercurialis annua (annual mercury) 43
Metazachlor (Butisan S) 28, 169, 170, 177, 197–201, 278
Midstream (diquat) 28, 280
Molinia caerulea (purple moor grass) 41, 85, 175
Mouse ear (*Cerastium* spp.) 49
MSS Atrazine 80WP *see* atrazine
MSS Atrazine 50FL *see* atrazine
MSS Atrazine 4G *see* atrazine
MSS 2,4-D *see* 2,4-D
Mugwort (*Artemesia vulgaris*) 49
Mustard, white 174
Mustard, black 174
Myosotis arvensis (forget-me-not) 47

Nettle, small 174
Nightshade, black 174
Norway maple 70

Oak (*Quercus* spp.) 70, 122
Onion, couch 174
Orache, common 174
Outlaw *see* Glyphosate

Pale persicaria (*Polygonum lapathifolium*) 49, 174
Pansy, field 174
Pansy, wild 174
Papaver rhoeas (corn poppy) 45
Parable (diquat with paraquat) 35, 37, 277

Paraquat (Dextrone X, Gramoxone, Scythe) 28, 278, 279
Paraquat with diquat (Farmon, Parable) 35, 37, 277
Parsley, piert (*Aphanes arvensis*) 49, 174
Paucus carota (wild carrot) 52, 176
Pearlwort (*Sagina* spp.) 49
Pendimethalin (Stomp) 28, 169, 170, 202–5, 278
Perennial/stinging/common nettle (*Urtica dioica*) 49
Perennial sow-thistle (*Sonchus arvensis*) 49
Picea abies (Norway spruce) 59, 60, 70, 75, 77, 83, 87, 89, 94, 124, 130, 134, 171
Picea omorika (Serbian spruce) 77, 124
Picea sitchensis (Sitka spruce) 70, 75, 77, 83, 87, 94, 124, 130, 134, 135, 136, 171
Pimpernel, scarlet (*Anagallis arvensis*) 174
Pine, Corsican (*Pinus nigra* var. *maritima*) 70, 75, 77, 83, 87, 124, 130, 134, 135, 171
Pine, lodgepole (*Pinus contorta*) 75, 77, 83, 87, 124, 130, 134, 135
Pine, Scots (*Pinus sylvestris*) 70, 75, 77, 83, 89, 124, 130, 171
Pineapple weed (*Chamomilla suaveolens*) 49, 174
Pinus contorta (lodgepole pine) 75, 77, 83, 87, 124, 130, 134, 135
Pinus nigra var *maritima* (Corsican pine) 70, 75, 77, 83, 87, 124, 130, 134, 135, 171
Pinus sylvestris (Scots pine) 70, 75, 77, 83, 89, 124, 130, 171
Plantago spp. (plantain spp.) 50
Plantago lanceolata (ribwort plantain) 50
Plantago major (greater plantain) 47
Plantago media (hoary plantain) 48
Plantain spp. (*Plantago* spp.) 49
Poa annua (annual meadow grass) 42
Poa pratensis (smooth meadow grass) 42
Poa trivialis (rough meadow grass) 42
Polygonum aviculare (knotgrass) 48
Polygonum japonicum (Japanese knotweed) 123
Polygonum lapathifolium (pale persicaria) 49, 174
Polygonum persicaria (redshank) 50, 175
Poplar 70
Poppy, common 174
Populus spp. (aspen poplar) 122
Portman Glyphosate *see* Glyphosate
Potentilla repens (creeping cinquefoil) 45
Privet (*Ligustrum vulgare*) 122
Propaquizafop (Falcon 100, Shogun 100 EC) 28, 170, 171, 206–7, 278
Propyzamide (Kerb 50W, Kerb Flowable, Kerb Granules) 28, 35, 36, 37, 91-3, 169, 170, 177, 258, 278

Prunella vulgaris (self heal) 50
Prunus laurocerasus (laurel) 122
Prunus spinosa (blackthorn) 122, 153
Pseudotsuga menziesii (Douglas fir) 70, 75, 77, 83, 87, 89, 94, 124, 130, 134, 136, 171
Purple moor grass (*Molinia caerulea*) 41, 85, 175

Quercus spp. (oak) 70, 122

Radish, wild 175
Ragwort (*Senecio jacobaea*) 50
Ramex obtusifolius (broad dock) 43
Ranunculus spp. (buttercup spp.) 43
Ranunculus acris (meadow buttercup) 48
Ranunculus arvensis (corn buttercup) 44
Ranunculus repens (creeping buttercup) 45
Raphanus raphanistrum (wild radish) 51
Red deadnettle (*Lamium purpureum*) 50
Red fescue (*Festuca rubra*) 41
Redshank (*Polygonum persicaria*) 50, 175
Reglone (diquat) 28, 280
Regulox K (maleic hydrazide) 280
Rhamnus cathartica (buckthorn) 122
Rhododendron (*Rhododendron ponticum*) 122, 155–67
Rhododendron ponticum (rhododendron) 122, 155–67
Ribwort plantain (*Plantago lanceolata*) 50
Rival (glyphosate) 279
Root-out (ammonium sulphamate) 28, 120, 139–40, 146–7, 155, 156–7, 240, 277
Rosa canina (wild rose) 122
Rosebay willowherb 175
Rough meadow grass (*Poa trivialis*) 42
Roundup Biactive *see* Glyphosate
Roundup Pro Biactive *see* Glyphosate
Rowan (*Sorbus aucuparia*) 122
Rubus spp. (bramble) 122
Rumex spp. (dock/sorrel spp.) 46, 51
Rumex crispus (curled dock) 46
Rush (*Juncus* spp.) 42
Rye grasses 175

Sagina spp. (pearlwort) 49
Salix spp. (willow) 122

Sambucus nigra (elder) 122
Scarlet pimpernel (*Anagallis arvensis*) 50
Scented mayweed (*Chamomilla recutita*) 50, 174
Scentless mayweed (*Matricaria perforata*) 50, 174
Scythe (paraquat) 28, 278, 279
Sedges 175
Self heal (*Prunella vulgaris*) 50
Senecio jacobaea (ragwort) 50
Senecio vulgaris (groundsel) 47
Sheep's fescue (*Festuca ovina*) 40
Sheep's sorrel 175
Shepherd's purse (*Capsella bursa-pastoris*) 50, 175
Silvapron D(2,4-D) *see* 2,4-D
Sinapis arvensis (charlock) 43, 172
Small nettle (*Urtica urens*) 43
Small reed grass (*Calamagrostis epiqejos*) 38, 85, 93
Smooth hawksbeard (*Crepis capillaris*) 51
Smooth meadow grass (*Poa pratensis*) 42
Smooth sow-thistle (*Sonchus oleraceus*) 51
Soft broom 175
Solanum nigrum (black nightshade) 43
Solidago canadensis (golden rod) 47
Sonchus arvensis (perennial sow-thistle) 49
Sonchus asper (annual milk or sow thistle) 43
Sonchus oleraceus (smooth sow thistle) 51
Sorbus aucuparia (rowan) 122
Sorrel spp. (*Rumex* spp.) 51
Sow thistle (annual milk thistle) (*Sonchus asper*) 43
Spear thistle (*Cirsium vulgare*) 51
Speedwell, common 175
Speedwell, germander 175
Speedwell, grey 175
Speedwell, green 175
Speedwell, ivy-leaved 175
Speedwell, wall 175
Spergula arvensis (corn spurrey) 45
Spruce, Norway (*Picea abies*) 59, 60, 70, 75, 77, 83, 87, 89, 94, 124, 130, 134, 136, 171
Spruce, Serbian (*Picea omorika*) 77, 124
Spruce, Sitka (*Picea sitchensis*) 70, 75, 77, 83, 87, 94, 124, 130, 134, 135, 171
Spurge (*Euphorbia* spp.) 51
Spurrey, common 175

St John's wort (*Hypercium perforatum*) 51
Stefes Glyphosate *see* Glyphosate
Stefes Kickdown 2 *see* Glyphosate
Stellaria media (common chickweed) 44
Stetson *see* Glycophosate
Stinking chamomile (*Anthemis* spp.) 51
Stirrup *see* Glyphosate
Stomp (pendimethalin) 28, 169, 170, 202–5, 278
Sweet chestnut 70
Sweet vernal (*Anthoxanthum odoratum*) 38, 175
Sycamore (*Acer* spp.) 70, 122

Tall fescue (*Festuca arundinaceae*) 40
Taraxacum officinalis (common dandelion) 44
Tare spp. (*Vicid* spp.) 51
Terbuthylazine with atrazine (Gardoprim A 500FW) 279
Terbutryn (Claroson 1FG) 280
Thale cress (*Arabidopsis thaliana*) 51
Thistle, creeping 175
Thistle, smooth sow 175
Thistle, spear 175
Thlaspi arvense (field penny-cress) 46
Thuja plicata (western red cedar) 70, 77, 83, 94–6, 124, 130, 171
Timbrel (triclopyr) 28, 35, 36, 37, 94–6, 120, 121, 136–8, 144–5, 153–4, 155, 166–7, 278
Timothy 176
Tracker (dicamba) 28, 97, 101–2, 277, 279
Trefoiles (from seed) (*Trifolium*) 52
Triclopyr (Chipan Garlon 4, Garlon 4, Timbrel) 28, 35, 36, 37, 94–6, 120, 121, 136–8, 144–5, 153–4, 155, 166–7, 278
Triclopyr with 2,4-D and dicamba (Broadshot) 35, 36, 37, 79–80, 120, 121, 126–7, 148–9, 155, 158–9, 277, 280
Trifolium spp. (clover spp.) 44
Trifolium repens (white clover) 52
Tripart Ratio (isoxaben) 28, 35, 36, 37, 89–90, 169, 170, 174, 278, 279
Tsuga heterophylla (western hemlock) 59, 77, 124
Tufted hair grass (*Deschampsia caespitosa*) 39, 176
Tussilago farfara (coltsfoot) 44
2,4-D (Atlas 2,4-D, BH 2,4-D Ester 50, Dicotex Extra, Dormone, MSS 2,4-D, Silvapron D) 25, 28, 35, 36, 37, 77–8, 109, 111–13, 120, 124–5, 277, 279
2,4-D with dicamba and triclopyr (Broadhot) 35, 36, 37, 79–80, 120, 121, 126–7, 148–9, 155, 158–9, 277, 280
2,4-D Ester 50 25, 28, 35, 36, 37, 77–8, 109, 111–13, 120, 124–5, 277, 279

Ulex spp. (gorse) 122
Ulmus procera (elm) 122
Unicrop Flowable (atrazine) 28, 35, 36, 37, 59–69, 277, 279
Urtica dioica (perennial/stinging/common nettle) 49
Urtica urens (annual/small nettle) 43

Velpar (hexazinone) 229
Veronica hederifolia (ivy-leaved speedwell) 48
Veronica persicae (field speedwell) 46
Vetches (from seed) (*Vicia* spp.) 52, 176
Viburnum opulus (guelder rose) 123
Vicia spp. (vetches) 52
Vicia sativa (common vetch) 44
Vicid spp. (tare spp.) 51
Viola arvensis (field pansy) 46
Viola tricolor (wild pansy) 52
Viper's bugle (*Ajuga* spp.) 52
Volunteer cereals 176
Volunteer oilseed rape (*Brassica napus*) 52, 176
Volunteered (Di-1-menthane with dalapon) 279

Wavy hair grass (*Deschampsia flexuosa*) 40, 85
White clover (*Trifolium repens*) 52
Wild carrot (*Paucus carota*) 52, 176
Wild cherry 70
Wild oat 176
Wild pansy (*Viola tricolor*) 52
Wild radish (*Raphanus raphanistrum*) 51
Wild rose (*Rosa canina*) 122
Willow (*Salix* spp.) 70, 122
Willowherb (*Chamerion angustifolium*) 52
Wood small reed 176
Woody weeds 120–54

Yarrow (*Achillea millefolium*) 52
Yellow oat grass 176
Yorkshire fog (*Holcus lanatus*) 41, 85, 176

Printed in the United Kingdom for HMSO
Dd 297424 C50 6/95 552 12521